Philosophical Foundations of Probability Theory

International Library of Philosophy

Editor : Ted Honderich

A catalogue of books already published in the
International Library of Philosophy
will be found at the end of this volume

Philosophical Foundations of Probability Theory

Roy Weatherford

University of South Florida

ROUTLEDGE & KEGAN PAUL

London, Boston, Melbourne and Henley

First published in 1982
by Routledge & Kegan Paul Ltd
39 Store Street, London WC1E 7DD,
9 Park Street, Boston, Mass. 02108, USA,
Broadway House, Newtown Road,
296 Beaconsfield Parade,
Middle Park,
Melbourne, 3206,
Australia and
Henley-on-Thames, Oxon RG9 1EN
Set in Times Roman by
Thomson Press (India) Ltd
and printed in Great Britain by
Thomson Litho Ltd
East Kilbride, Scotland
© Roy Weatherford 1982

Library of Congress Cataloging in Publication Data

Weatherford, Roy, 1943-
Philosophical foundations of probability theory.

(International library of philosophy)
Bibliography: p.
Includes index.
1. Probabilities. I. Title. II. Series.
BC141.W4   121'.63   81-22730

ISBN 0-7100-9002-1   AACR2

For

DORIS
Wife, lover, and friend

CONTENTS

Contents

PREFACE

When I was writing my dissertation on C. I. Lewis's epistemology, I was struck by his epigram 'If anything is to be probable, then something must be certain' (*An Analysis of Knowledge and Valuation*, p. 186).

If Lewis were right about this, it seemed to me, then epistemology must indeed take the form of a Cartesian reconstruction, seeking to base our merely probable empirical knowledge on some foundation of ultimate certainties. And yet Lewis obviously thought that this principle, with its profound philosophical implications, could readily be established by merely mathematical considerations from the theory of probability. While I consider myself to be as ardent an admirer of science and mathematics as any good product of the American school of hard analytic philosophy, I found it hard to believe that the very structure of human knowledge could be dictated by relatively trivial mathematical theorems. There must, I thought, be something deeper and more philosophical in probability theory than I had learned in the course of taking my bachelor's degree in mathematics. And so I walked over to Widener library to check out a good book on the specifically philosophical aspects of probability theory – and I couldn't find one!

I found many good books on the mathematics of probability theory, a few on its history, and a handful which discussed the various 'interpretations' of probability, but none which gave a comprehensive discussion of the metaphysical and epistemological roots and branches which were important to me. In frustration I resolved to write the book I could not find.

The project has been longer and more arduous than I would have wished. I quickly decided that no one interpretation was sufficiently dominant to be my single subject and that, indeed, it is very important to see how the different theories have varying philosophical assumptions and implications. The result is that I had to attempt to understand and analyze not just one, but at least four different theories of probability.

In addition to this expansion of my topic, I also discovered that my personal commitments were expanding in a way which made it difficult to find research time. My university expects faculty to distinguish themselves in the three fields of teaching, research, and service. Unfortunately, I have found myself unable to do all three – or even two – simultaneously and well. In teaching, I am a compulsive preparer of lectures, so that when I teach even one or two courses my preparation expands to fill all available time. In service, I am the chief lobbyist for my union, the United Faculty of Florida, American Federation of Teachers, AFL-CIO, and find it necessary to devote an entire quarter each year to the Florida legislative session – an enjoyable and stimulating activity but one which leaves little time for reading and writing. And finally, for research, I have found that my personal style of scholarship requires at least an entire day of effort, and preferably weeks and months of total immersion, before I can get to the stage of putting down even a sentence or two. The result is that a research project which should have been finished in a couple of years has stretched out longer than I care to remember. That I was able to complete it at all is entirely due to the support I have received from others to enable me over the years to take a summer here, a quarter there, to devote entirely to writing.

Of those who have made this possible I must thank first the University of South Florida, and especially President John Lott Brown, Dean David Smith, and Philosophy Chairperson Willis Truitt. Their personal kindness and concern for scholarship resulted in my getting far more encouragement, support, and release time than most union activists expect or receive.

Among foundations, the Danforth Foundation supported me as a Danforth Fellow during those early years at Harvard when I encountered this problem and read the basic texts. The National Endowment for the Humanities, through its program of Summer Seminars for College Teachers provided me with two very enjoyable and productive summers, one with Roderick Chisholm at Brown

University and the other with Richard Rorty at Princeton. The final chapter and editorial revisions were completed under a grant from the University of South Florida Foundation, for which I am very grateful.

My philosophical debts are too numerous to recount, as I seem to have been working on this project for most of my philosophical life. I must, however, mention at least Donald St. Clair, who first got me addicted to philosophy at Arkansas Tech; Roderick Firth, Israel Scheffler, Rogers Albritton, John Rawls, and (especially) Hilary Putnam of the Harvard faculty, all of whom have belied the claim that eminent scholars care only for their own work and have no time for students and younger philosophers; Bob Schultz, Paul Gomberg, and Howard Rolston who taught me how to 'talk philosophy' as graduate students; and my colleagues at the University of South Florida – especially Willis Truitt and Bruce Silver – who have helped me 'do philosophy' as a professor.

From other fields, I am grateful for the assistance of David Stroud in physics and Sandy Turner in mathematics.

Ted Honderich, the general editor of this series, was the source of useful practical advice and much-needed psychic support and encouragement. David Godwin and the editorial staff were kind and helpful in leading me through the unfamiliar complexities of preparing a manuscript for publication.

Despite the help of all these good people and institutions, there were months when no income was available, days when it scarcely seemed worth it to continue, and years and years of unreasonable demands on my family. For putting up with all this, and for helping me through it, my wife Doris and my daughter Meg deserve the greatest share of my love and appreciation.

I

WHAT IS PROBABILITY?

Style manuals advise us that the proper way to begin a piece of expository writing is to introduce and identify clearly the subject of our exposition. This would seem, then, to be the appropriate place to offer a precise definition of 'probability.' Unfortunately, we are unable to do so. In fact, one of the major disputes of probability theorists is precisely the question of what is to count as an appropriate definition of probability; so that if we were to begin this chapter by arbitrarily deciding that crucial issue, we should be, like the White Queen, living backwards and arriving at our conclusion before we conduct our investigation. We shall therefore postpone the question of definition, and indicate instead some general outlines of our use of the concept, hoping thereby to gain some idea of what we are talking about when we talk about probability.

In daily life we find more frequent use of the adverbial form 'probably' or the adjectival form 'probable' than of the substantive 'probability.' This is presumably because ordinary conversation generally *employs* abstract or quantitative ideas to talk *about* concrete physical objects or events. Thus, we are more likely to say Jones will probably win the election, than 'The probability that Jones will win the election is high,' for about the same reasons that we are more likely to say 'There are three apples on the table' than to say 'The number of apples on the table is three.' The reason is that we normally talk about apples instead of numbers and people instead of probabilities.

There are many different ways in which we use these terms in ordinary life, and many different propositions and substantives which

1

we modify. In this general class of examples we find such usages as:

Caesar probably visited Britain.
The outbreak of a nuclear war is less probable now than it was
10 or 15 years ago.
The likely winner is Miss Florida.
The expanding universe theory is probably true.
The door is probably locked.

Even when we fail to use such terms as 'probable' or 'likely' in a sentence, we often consider them as implicit qualifiers and will add them on request:

'Taxes will go up again next year.'
'That's not certain.'
'Maybe not, but it's damn sure probable.'

This example is an instance of *prediction*. Human beings have long realized that it is difficult to make *accurate* statements about the future and virtually impossible to make *certain* ones. One of the general guidelines on the use of 'probability' that has long been recognized is therefore that probability has to do with *predicting the future*.

Yet by looking back at our list of examples above, we see that the Caesar example concerns the past and the door example concerns the present, while the expanding universe example is apparently timeless. Thus it is not the case that probability is *always* concerned with the future. These examples are analogous to the tax dialogue not in being predictive, but in concerning an assertion that is less than certain. Perhaps, then, we can say that at least some uses of probability involve *asserting with less than certainty*.

When we assert something of which we are uncertain, we usually do so because we have some evidence which supports the assertion although it is not conclusive. For example,

Since Arkansas beat Texas, they will probably be the Southwest
Conference champions.
On the evidence presented to the inquest, I find that the
probable cause of death was murder.
Data from Apollo 15 make it more probable that the Moon
has experienced vulcanism.

Such usages as these suggest that probability concerns the *support*

2

of a conclusion by given evidence or the relationship between premises and conclusion in a non-deductive argument.

Just as we do sometimes talk about numbers rather than things, it is also true that we sometimes talk about probabilities rather than events. Hence,

> The probability of rain tomorrow is 70 per cent.
> The odds on Gluefoot are 20 to 7.
> His chance of failing is greater than one-half.
> The probability that the Moon was formed in its present orbit is greater than the probability that it was captured by the Earth.

In such contexts it becomes clear that at least part of the use of the term requires that 'probability' refers to an *abstract quantity or number*.

Finally, when we inquire into the nature and origin of such numbers, we find that they often are based on the frequency with which a property appears in a population. Thus,

> Since 50 per cent of inductees become combat soldiers and 20 per cent of combat soldiers become casualties, an inductee has a 10 per cent chance of becoming a casualty.
> Since 4 of the 52 cards are Aces, the probability of drawing an Ace is 1/13.
> Of a thousand men in your age group, 50 die in a given year: thus your probability of surviving the year is 0.95.

These last examples suggest the possibility that a probability is a *frequency ratio in a population*.

All of these guidelines, restrictions, and rules of usage have some bearing on our use of the word 'probability,' and must at least be considered in any theory of probability if that theory is intended to explicate our ordinary usage as well as regulate it.

C. I. Lewis has said that he could not *explain* probability to anyone who did not already possess a primordial *sense* of probability.[1] John Maynard Keynes has said, in a similar vein, 'A *definition* of probability is not possible, unless it contents us to define degrees of the probability-relation by reference to degrees of rational belief. We cannot analyze the probability-relation in terms of simpler ideas.'[2] Others have held that a definition is possible, but only in the context of developing a theory of probability. Considering the difficulties we have met so far, let us see, then, if we can make more progress by inquiring about theories of probability rather than definitions *simpliciter*.

1   WHAT IS A THEORY OF PROBABILITY?

If we consider the theory of probability to be in the same boat as most other more or less scientific theories, we shall find it to be a philosophically rocky boat indeed, adrift in an uncharted sea. For questions about the nature, origin, and justification of theories are among the most widely debated issues in Philosophy of Science today. It would truly be presumptuous of us to attempt to lay down strict general rules to which any theory must accord; yet this need not inhibit us from trying to say a few things about what theories of probability have traditionally had in common and how they have differed and what it would be reasonable to require of future theories.

We mentioned above that a theory must at least *consider* the various conceptual outlines and linguistic usages which are sanctioned by either scientific or ordinary discourse, and it would seem appropriate to require that it either *incorporate* or at least *account for* as many of these pretheoretical instances as possible.[3] This is so because otherwise it would not be a theory of probability, but of something else. Ideally, the theory should account for *all* uses of 'probability', but that is far too restrictive a condition to impose.

Despite the fact that it must be *based on* the received, standard use of 'probability,' the avowed aim of a theory of probability is to *sanction and regulate* such uses. If it were *not* thus prescriptive, it would not be a *theory* of probability, but at best a description of current usage. It seems, then, that valid instances of probability judgments must agree with our theory, and that the theory is justified by its conformity to valid probability judgments.[4] This mutual accommodation between a theory and its subject matter is developed in Nelson Goodman's discussion of inductive inference:[5]

> The task of formulating rules that define the difference between valid and invalid inductive inferences is much like the task of defining any term with an established usage. If we set out to define the term 'tree', we try to compose out of already understood words an expression that will apply to the familiar objects that standard usage calls trees. A proposal that plainly

violates either condition is rejected; while a definition that meets these tests may be adopted and used to decide cases that are not already settled by actual usage. Thus the interplay we observed between rules of induction and particular inductive inferences is simply an instance of this characteristic dual adjustment between definition and usage, whereby the usage informs the definition, which in turn guides extension of the usage.

Following this principle, we find that 'the familiar objects that standard usage calls' probability judgments include above all those strictly mathematical relations and manipulations which have come to be known as the *calculus of probability*. This calculus is a system for manipulating numbers of a certain type in order to produce more numbers of the same type. The calculus was developed piecemeal, largely in an attempt to calculate the odds in various games of chance. Its general principles have become fairly standard in theories of probability, presumably because they capture well our intuitions about the relations between probabilities. Despite their historical origin as empirical descriptions of games of chance, Reichenbach[6] and others have shown that such rules as the multiplication law for independent events and the addition law for exclusive alternatives can be developed rigorously and consistently as an uninterpreted axiom system; it is then up to each theory to give an *interpretation* to the calculus which will lead to our familiar probability judgments.

Whether the calculus derives from experience or is established axiomatically or results from a particular definition of 'probability', the strictest requirement for a theory of probability is that it must contain such a calculus in a form which varies little from the standard one.

Given this start, it is obvious that the second requirement is that a theory must give an *interpretation* of the calculus, or a definition of 'probability,' which will enable us to use that calculus in making most of the probability judgments which we do in fact make.[7]

We shall now examine the most important attempts to fulfill these requirements with a theory of probability.

2 TYPES OF PROBABILITY THEORIES

In this book we will examine four major types of probability theory. Some authorities recognize more: I have seen up to eleven different senses of 'probability' distinguished by an analyst. Most, however, recognize fewer. Carnap and Lewis agree there are only two, while Nagel accepts three. (These different schemes of classification will be discussed in the next section, after the reader is familiar with the basic types.)

These are the four theories with which we will be concerned:

1 The Classical Theory of Probability (CTP): defines probability in terms of ratios of equipossible alternatives.
2 The A Priori (AP) Theory: defines probability as a measure of the logical support for a proposition on given evidence.
3 The Relative Frequency (RF) Theory: defines probability as the (limit of the) relative frequency of appearance of one (infinite) class in another.
4 The Subjectivistic Theory (SUB): defines probability as the degree of belief of a given person in a given proposition at a specific time.

Certainly one could make a case for dealing with more (or fewer) theories.

There are two major reasons why I have chosen not to include more theories: (1) some, such as Braithwaite's, Popper's, and Toulmin's, are not (yet) important enough and have not been accepted and used by sufficient numbers of philosophers, mathematicians, and statisticians; (2) some, such as Wittgenstein's and Lewis's, are not distinct enough in their identity but share most features with a more important theory which is discussed here.

On the other hand, I have expanded my coverage beyond the two (AP and RF) recommended by Carnap (*inter alia*) because: (1) The Classical theory retains several differences from its descendant, the AP theory, and it is also of sufficient historical importance to warrant its inclusion in a work of this magnitude. (2) The Subjectivistic theory has grown in importance since Carnap's major work was written and has become the preferred view of many (especially Bayesian) statisticians and the working model of many experimental psychologists and decision theorists.

On the whole my selection has been made on the grounds that these are the most important theories, both historically and theoretically. To delete one of them would seriously truncate our subject; to add on another would not be proportionately valuable.

Now let us take a quick introductory overview of the kinds of theories and establish their major differences.

The Classical Theory of Probability (CTP)

The 'Classical Theory of Probability' is a slightly artificial name for the views of the founders of the theory of probability. Such men as Jacob Bernoulli and Pierre Simon de Laplace were generally not developers and adherents of an articulated and self-consistent formal system of the type we would today call a theory. Rather they shared a common approach and some basic ideas as they constructed the mathematical heart of probability theory – the probability calculus.

For the most part, I have treated this school as a historical entity, rather than a current contender for theoretical respectability. To that extent the adherents of the CTP are identifiable chronologically, as participants in a certain period of the development of our subject. Yet there are issues and attitudes involved in the CTP. The body of doctrine which constitutes roughly the core of this somewhat unstructured theory goes something like this:

(a) Probability is the ratio of the number of favorable cases to the number of all equipossible cases.

(b) The Principle of Indifference: Events are equipossible if we have no reason to prefer or expect one over the other (later: if they are coordinate events, symmetrically related to the evidence).

(c) There is no objective chance or indeterminism – probability is a measure of our partial ignorance.

(d) Nevertheless, there are objective rules for generating and combining probabilities – it is not just a matter of opinion.

(e) Repetitive events with fixed probabilities have an expected frequency of occurrence (Bernoulli's Theorem). It may be possible to use an observed frequency to infer the fixed but unknown probabilities of some events (Bayes's Theorem, Inverse Bernoulli).

To this day, most gamblers and ordinary folks rely on the CTP for those few types of quantitative probability judgments most of us make. This is how we know that the probability of an ace is 1/13 and that we shouldn't draw to an inside straight. But theory has marched on, and now most advanced probabilists prefer one of the other theories of probability.

The A Priori (AP) Theory of Probability

A theory of probability will be said to be of the a priori type if most of the following conditions are met:

(a) It describes probability as a logical relation between statements.
(b) It considers this relation to be completely determinable by the application of logic and the rules of probability to the two statements.
(c) It requires that every ascription of probability must be relative to certain evidence. Unqualified ascriptions of probability are either elliptical or meaningless.
(d) It considers every properly derived probability statement to be analytic, logically true, and incorrigible.
(e) It holds that 're-evaluation' or 'correction' of a probability statement consists actually of its replacement by another probability statement which is also logically true but refers to different evidence and therefore ascribes a different value to the probability relation.

To a great extent, the a priori theory is the inheritor of the tradition of the Classical Theory of Probability.[8] Most notoriously, AP theories tend to incorporate some form of the Principle of Indifference into their structure. This makes it possible for them to generate initial probabilities without the laborious empirical investigations required by relative frequency (RF) theories. It also, of course, subjects them to the same charges of hocus-pocus and 'making knowledge out of ignorance' which have bedeviled the CTP.

AP theorists commonly reply that their discipline is a part of logic, showing the evidentiary connection between what we know and what we can predict. Seen in this light, they say, their theories tell us no more and no less about reality than do formal mathematics and deductive logic, which are also a priori.

In later years, under the influence of Rudolf Carnap, AP theorists have spent much of their time seeking to develop a formal inductive logic based on the structure of language.

8

Relative Frequency (RF) Theories of Probability

During the middle part of this century the principal opponent of AP probability theory has been the view that probability is not some logical, abstract connection between words and sentences but rather is the actual, empirical rate of occurrence of some feature of the real world. This view identifies the *probability* of occurrence of X with the *actual* relative frequency of occurrence of X in the real world, and was developed largely by Richard von Mises and Hans Reichenbach and has been especially attractive to scientists and actuaries. Generally, a theory is said to be of the relative frequency (RF) type if most of the following conditions are met:

(a) It defines 'probability' as 'the relative frequency of a property within a population.'
(b) It defines 'probability' (in at least some cases) as 'the limit of a relative frequency in an infinite series.'
(c) It holds that the probability calculus is an axiomatic mathematical tool for dealing with reality – just as arithmetic and geometry are.
(d) It holds that individual probability-statements attribute an empirical property to an empirical population.
(e) It holds, therefore, that although theorems of the probability calculus may be analytic, individual probability-statements are definitely synthetic, empirical, and factual.

The fundamental difference between the RF and AP theories is that RF probability is an empirical, measurable property of the actual physical world, while AP probability is a formal, logical property of the way we think and speak about the world.

The Subjectivistic Theory (SUB) of Probability

Our final theory of probability is the newcomer to the group. Its theoretical development began with Frank P. Ramsey and was largely completed by Bruno de Finetti and Leonard J. Savage. The principal views of this school are as follows:

(a) Probability is the degree of belief of a given person in a given proposition.

(b) Probabilities are best established by examining behavior, especially betting behavior.

(c) There are no objective probabilities – or at least this is a different and less important sense of 'probability.'

(d) An event has no unique probability. Each individual is logically free to set his own values.

(e) The probability beliefs of a rational individual must be consistent and governed by the Calculus of Probability.

This view of probability turns essentially on the thesis that each person's probability beliefs are ineliminably private, personal, and subjective. There cannot be rules (according to SUB) which will tell me how strongly to *expect* the occurrence of X any more than there can be rules telling me how much to *fear* the occurrence of X. Both are subjective attitudes, and none of us has the ability or the right to tell anyone else what to fear or expect in normal cases. In abnormal cases, however, we may have need of therapy.

Psychological therapy is needed when our fears become excessive, incapacitating, neurotic. Probability therapy is needed when our expectations are inconsistent (and therefore irrational).

If we allow ourselves to believe and act upon inconsistent probability assessments, we become the losers in an operation known as the Dutch Book (a series of bets on an event, at varying odds, which guarantee one side a net profit and the other a net loss). By studying probability theory we can learn to avoid being victimized in this manner.

It is the peculiar view of de Finetti that the only definite knowledge available from probability theory is a form of the Principle of Non-contradiction. Once we understand the calculus and know how to keep our beliefs consistent, probability theory can tell us nothing more about *what* to believe. It cannot tell us, for example, that the probability that this card is an ace is 1/13, because there are no rules for determining such a probability and, in fact, there is no such objective probability at all to be discovered. Probabilistic reasoning is more an art or skill than a science, and the best we can hope for is some Counsels of Prudence or Rules of Thumb about reasonable ways of arriving at a probability value.

This is such a truncation of the scope of probability theory that many who study SUB have refused to follow de Finetti in this view. Those who work in experimental psychology or decision theory, for example, often hold that subjectivistic probabilities are one interesting phenomenon but are far from exhausting probability theory. Many continue to believe in the existence of objective probabilities (whether AP or RF) and in the importance of identifying them as well as discovering how individuals do behave. But the official position of SUB theorists is that all probability phenomena are personal, subjectivistic, degrees of belief or expectation.

3 SCHEMES OF CLASSIFICATION OF THEORIES

Now that we are familiar with the four basic types of theories to be discussed in this book, let us digress a moment to consider other ways in which our subject might have been arranged.

The classification of theories of probability is, like most taxonomies, a matter of taste as well as fact, convenience as well as correctness. Natural kinds do seem to exist among the theories, but their differentia are neither physically nor logically determinate. Rather, we have several individual theories characterized by Wittgensteinian 'family resemblances.' One must decide which features are most important for speciation as well as which theories 'really do' possess which features before one can present a completed classification. These decisions are certainly debatable and somewhat subjective.

Yet a classification is necessary. There are too many individual theories – even too many *major* theories – for the average student of the subject to attempt to master them all. And taxonomies are not just an intellectual economy measure. In stressing similarities and identifying differences the taxonomist is forced to decide and dramatize *which features are important* in his field; similar decisions on a smaller scale must be made about each theory as it is classified. The resulting system, if it is well thought out, can be useful to the expert as well as the beginner by identifying contrasts and comparisons, consistencies and inconsistencies which might otherwise have gone unremarked.

It might be fun, and even instructive, to attempt a taxonomy of taxonomies, showing how different critics have been influenced by the same and different considerations in drawing their categories of probability theories. Rather than enter upon this topic of second-order criticism, I will complete this preliminary chapter by an

11

unsystematic description of other classification schemes colored by a few remarks in support of my own.

Carnap's Grand Dichotomy

Obviously the simplest way to differentiate a topic is to break it up into two parts. Many writers have done this to probability theory, though the break is not always found in the same place.

The most famous of these twofold divisions is that of Rudolf Carnap, who grouped theories according to the underlying concepts they sought to explain:[9]

> The various theories of probability are attempts at an explication of what is regarded as the prescientific concept of probability. In fact, however, there are two fundamentally different concepts for which the term 'probability' is in general use. The two concepts are as follows, here distinguished by subscripts.
>
> (i) Probability$_1$ is the degree of confirmation of a hypothesis *h* with respect to an evidence statement *e*, e.g., an observational report. This is a logical semantical concept. A sentence about this concept is based, not on observation of facts, but on a logical analysis; if it is true, it is L-true (analytic).
>
> (ii) Probability$_2$ is the relative frequency (in the long run) of one property of events or things with respect to another. A sentence about this concept is factual, empirical.

This division is simple and plausible and recurs in one form or another in many theorists (C. I. Lewis,[10] for example). This distinction roughly parallels that between our A Priori (AP) and Relative Frequency (RF) groupings. In my doctoral dissertation, I too accepted this grand dichotomy and used it in my discussion of theories of probability.[11] I have abandoned it in this work for three principal reasons:

1 A more extended and detailed discussion invites and permits the treatment of more major types of theories.
2 Although the Classical Theory of Probability (CTP) resembles the AP theory in many respects, it lacks the fundamental insight that probability is a matter of logic rather

than the world, while retaining a historical and
pedagogical importance of its own.

3 The Subjectivistic Theory of Probability (SUB) has risen to
importance since Carnap's time and doesn't fit neatly into
either of his categories.

Nagel's Threefold Division

The *International Encyclopedia of Unified Science* was part of the
logical empiricists' grand design to bring system and order to the
scientific search for knowledge. One of its major articles was Ernest
Nagel's monograph, *Principles of the Theory of Probability*.[12] In
this work Nagel identifies three major types of theories of probability,
which correspond to 'three major interpretations' of the term
'probability.'[13]

1 According to the first, a degree of probability measures our
subjective expectation or strength of belief, and the calculus
of probability is a branch of combinatorial analysis; this is
the classical view of the subject, which was held by Laplace
and is still professed by many mathematicians. It is not
always clear whether by 'expectation' proponents of this view
understand *actual* expectations or *reasonable* expectations.

2 According to the second, probability is a unique logical
relation between propositions, analogous to the relation of
deducibility; its most prominent contemporary supporter is
the economist Keynes.

3 According to the third, a degree of probability is the measure
of the relative frequency with which a property occurs in a
specified class of elements; this view already appears in
Aristotle, was proposed by Bolzano and Cournot during the
last century and further developed by Ellis, Venn, and Peirce,
and was finally made the basis for a subtle mathematical
treatment of the subject by von Mises and other contemporary writers.

This system has the advantage of extending Carnap's simple
division by incorporating a third category. Unfortunately, Nagel has
included in this group the characteristics of both our CTP and SUB
classes of theories. I think there are good reasons for singling out

'the classical view of the subject' but they are not the reasons Nagel gives; and I think there are theories according to which probability is a measure of 'our subjective expectation or belief' but they are the later theories of Ramsey, de Finetti, and Savage rather than the earlier theories of Laplace and company.[14]

Variant Classifications

While Carnap's and Nagel's views are the most prestigious, there have been other attempts to classify theories of probability. The more recent of these tend to include at least four groups of theories. I will touch briefly on some of the most important such classifications.

Kyburg – The prolific contemporary authority Henry E. Kyburg Jr has variously distinguished from three to five types of probability theories. In his introduction to *Studies in Subjective Probability* (1964) he notes that 'There are essentially three types of connection [between probabilities and the world] that have been proposed: the empirical, the logical, and the subjective.'[15]

In his later and more comprehensive *Probability and Inductive Logic*,[16] however, Kyburg has extended his scheme backwards to incorporate the classical interpretation and forward to add on a fifth view, his own 'Epistemological Interpretation of Probability'. The first four are essentially the same as mine (or, rather, mine are essentially the same as his; although I have not intentionally lifted directly from Kyburg or anyone else, I have certainly profited from reading his works) and go far towards establishing this as the standard scheme of classification. Whether or not his own theory will be justified as a fifth major type only time will decide.

Good – according to I. J. Good, 'Each application of a theory of probability is made by a communication system that has apparently purposive behavior.' These entities might be men, Martians, or machines. 'One point of the reference to machines is to emphasize that subjective probability need not be associated with metaphysical problems concerning mind.'[17]

Considering such entities and their relations to the world, Good distinguishes four different types of probability:[18]

    (i) Physical probability – exists irrespective of organisms.

(ii) Psychological probability – values inferred from the behavior of entities.

(iii) Subjective probability – 'psychological probability modified by the attempt to achieve consistency, when a theory of probability is used combined with mature judgment.'

(iv) Logical probability – the hypothetical subjective probability of an infinitely rational being.

One could plausibly interpret type (i) as being substantially the same as our RF, types (ii) and (iii) as representing two aspects of SUB, and type (iv) as covering the spheres of CTP and AP.

Good's distinctions have the virtue of identifying some different metaphysico-epistemological senses of the relation between probability, organisms, and the world.

Von Wright – In various works, G. H. von Wright has employed different methods of grouping and describing theories of probability. As an example, we can distinguish the following types of interpretations according to *The Logical Problem of Induction*:[19]

A The Frequency interpretation.

B The 'Spielraum' (roughly, 'range') interpretation, which breaks down into:

    1 The logical Spielraum theory, where probability is a ratio of truth-possibilities of a sentence;

    2 The empirical Spielraum theory, where certain atomic propositions are shown empirically to be equipossible.

C Probability as a 'Grundbegriff' (fundamental or *sui generis* concept). Keynes, Jeffreys, etc., deny that the concept of probability can be exhaustively defined by any of the above terms or any like them.

D Probability as degrees of belief.

These categories are similar to, but not readily reducible to, ours. The AP and CTP classes of theories are here intermingled and redivided in a variant fashion.

Black – In his *Encyclopedia of Philosophy* article Max Black suggests the following classification:[20]

1 Mathematical Dogmatism – No definition of 'probability' is possible or necessary; probabilists should confine themselves to dealing with the mathematical theory.

2 Classical Theory and the Principle of Indifference.
3 Logical Theories (Keynes, W. E. Johnson, Carnap, Harold Jeffreys).
4 Frequency Theories – 'probability is, in all cases, to be identified with some suitably defined relative frequency.'
5 Subjective Theories – A degree of 'rectified' confidence is identified with probability, and can vary from person to person with no imputation of fault.

Black's first category is useful if one's primary concern is to categorize probabilists descriptively. It is, however, useless for our purposes, as it represents not a separate view on the philosophical foundations of probability theory, but a refusal to construct such a view.

Otherwise this scheme is essentially identical to the one adopted here, and, since it appears in the influential *Encyclopedia*, we may hope that something like this will become the standard view.

Fine – Theories of Probability, by Terence L. Fine,[21] is a technical, mathematically oriented discussion of the major theories of probability. Since Fine is interested primarily in mathematical, rather than philosophical, foundations, he devotes most of his space to formal exposition and analysis of axiom systems and statistical techniques, with only a limited amount of philosophical analysis.

When he sets out to classify theories, Fine identifies the following types:[22]

1 Axiomatic comparative.
2 Kolmogorov's calculus.
3 The usual relative-frequency theory.
4 Von Mises's relative-frequency theory.
5 The Reichenbach-Salmon relative-frequency theory.
6 Solomonoff's complexity-based theory.
7 Laplace's classical theory.
8 Jayne's classical theory.
9 Koopman's comparative logical theory.
10 Carnap's logical theory.
11 The De Finetti-Savage subjective-personalistic theory.

Fine's list has the virtue of being 'more extensive' than ours. In some respects and for some purposes it is a fine thing for a list to be extensive, comprehensive, even complete. But obviously the limit in this direction would be simply to list each and every theory of probability. This would be useful as a catalog or a compendium, but for theoretical purposes would be at best a prolegomenon to the analysis and classification which makes the diverse views understandable.

In search of a clear, simple set of categories I preferred to combine Fine's 3, 4, and 5 into the one RF category, to skip 8 as historically and systematically unimportant, etc. Clearly this is not to dispute the validity of Fine's classification, but rather to adopt a different program to serve a different purpose.

4 CONCLUSION

For our purposes it is best to deal with four major types of theories: the Classical Theory of Probability (CTP), the A Priori (AP), the Relative Frequency (RF), and Subjectivistic (SUB) theories. This schema has the advantages of paralleling Kyburg's and Black's views, while usefully extending Carnap's and correcting Nagel's. There are theoretical and practical considerations which favor the adoption of this system, but the choice is largely one of heuristic convenience rather than ultimate theoretical import.

In what follows we will investigate these four types of theories in an effort to identify and describe clearly the philosophical foundations of each view. We shall be concerned especially with the metaphysical question of what probabilities *are* and the epistemological question of how probabilities *are known*. As usual, these issues are intertwined with the question of what probability statements *mean*.

We will try to begin with no preconceptions or desired findings. We will try to avoid imposing a grand unity on the one hand or multiplying niggling distinctions on the other. Perhaps at the end of our exploration we shall finally be able to answer our introductory question, 'What is Probability?'

II

THE CLASSICAL THEORY
OF PROBABILITY

The earliest attempts to deal with probability, from ancient times to the time of Laplace (1749–1827), are generally lumped together as the Classical Theory of Probability. This usage is perhaps slightly misleading, since it gives far too strong an impression that there existed some *theory*, some unified, consistent definition and explanation of probability to which the Classical theorists subscribed. One should not be surprised to find that this is not the case, since the relevant period includes only the beginning of probability theory and the first useful attempts to systematize it. Quite the contrary, one might well be surprised at the degree of agreement and systematic unity which *does* exist in the Classical writings. This agreement makes it possible to isolate a generally accepted definition of probability and some common ways of dealing with it. This body of thought was not clearly and systematically articulated by the Classical writers themselves (and philosophical analysis and exposition is especially lacking) but modern writers have generally agreed on its fundamental features so that one might say there *now* exists a Classical Theory of Probability while in Classical times there did not.

With these mild caveats in mind, then, we shall proceed in the usual manner to discuss the Classical Theory of Probability (CTP) and its adherents as though an explicit theory existed.

Since the CTP is normally attributed to a certain group of theorists, one might give an extensional definition of Classical Probability Theory by naming that group. Perhaps the simplest way of doing this is by saying that Classical theorists are those persons who appear in I. Todhunter's great work, *A History of the Mathematical Theory*

of Probability From the Time of Pascal to That of Laplace.[1]
This is a work that can truly be called monumental. It is large and thorough, and, as Keynes said, 'complete and exact, – a work of true learning, beyond criticism.'[2] It is also, however, dry, overly-technical, and somewhat tedious, so that F. N. David has been led to remark that 'Todhunter will rarely be read for pleasure, although always for profit, by anyone interested in probability theory.'[3]

Todhunter will be our basic text in this chapter and will frequently be cited rather than the original sources. I have adopted this course because (1) many of the original documents are either rare or untranslated and searching them out is a task for the historian of probability and inappropriate for this work, (2) the historical and theoretical significance of these authors depends not so much on the exact original language as on the public impact of a work – and this impact is reflected in Todhunter.

If any reader prefers a more scholarly history of the subject, I could do no better than to direct him or her to Todhunter – one who rejects the very notion of secondary sources is welcome to search out the originals as an exercise in scholarship. For our purposes, Todhunter will suffice.[4]

Now that I have disclaimed the title of historian, I will naturally proceed to give a historical account of the Classical Theory.

1 THE PREHISTORY OF PROBABILITY

It seems likely that probabilistic reasoning has existed almost as long as man has been a rational animal, but no one knows exactly when or how it began. As for the more explicit uses of probability in games of chance, the earliest probably appeared when savages began playing with the astragalus. The astragalus is the small bone just above the heel-bone (sometimes called 'hucklebone,' or, erroneously, 'knucklebone') in mammals and is important because it has four distinctive sides or faces upon which it may rest. It can therefore be used as a primitive – and very inexact – die for gaming. We do not know when it first came to be used for this purpose, but archaeologists have found astragali in disproportionate numbers in the campsites of prehistoric man. Whether or not it was used for gaming in those remote times, we do know that 'the astragalus was certainly in use for board games at the time of the First Dynasty in Egypt (*c.* 3500 B.C.),'[5] and gaming has been with us ever since.

Ancient Times

Gaming continued throughout the Ancient Greek and Roman era, so that Greek soldiers are reputed to have whiled away the tedious siege of Troy with various games of chance, and Roman emperors such as Augustus (63 B.C.–14 A.D.) and Claudius (10 B.C.–54 A.D.) are said to have been 'greatly devoted to dicing.'[6]

Furthermore, this era marks the beginning of enterprises relying on probability such as commercial and life insurance.[7] Still, these activities had not yet been put on a sound actuarial basis and no systematic discussion of probability or statistics has come down to us from this period.

Italian Renaissance

With the flowering of the Italian city-states and the rise of capitalism and commerce, commercial statistics and actuarial methods were developed for the first time.[8] It is also during this period that we find the first serious discussions of probability in the works of Fra Luca Pacioli (1445–1517?), Celio Calcagnini (1479–1541), and Tartaglèa (Nicola Fontana) (1500–57). However, these works were mainly descriptive and contained little of theoretical interest.[9]

Cardano

Girolamo Cardano (1501–76) occupies an ambiguous position in the history of probability theory. He was an inveterate gambler and a controversial man of learning (or plagiarism?) who wrote in or around 1525 a manuscript which was published posthumously in 1663 as *Liber de Ludo Aleae* (*Book on Games of Chance*). This book contains descriptions of certain games and the first published calculations of odds. It is disparaged by Montmort and Todhunter,[10] but praised by Libri and Maistrov.[11] David goes so far as to say that Cardano (or his assistant Ferrari) was the first 'to introduce the idea of combinations, to enumerate all the elements of the fundamental probability set, and to notice that if all the elements of this set are of equal weight, then the ratio of the number of favorable cases to the total number of cases gives a result in accordance with experience.'[12] To the extent that this is true, one might well describe Cardano as the father of the Classical Theory of Probability.

Galileo

Galileo Galilei (1564–1642) appears in our history for two reasons:

1. He gave the first *recorded* calculation of possible outcomes with three dice, and
2. He made some important general remarks on the theory of errors.

However, in the former matter he appears to have functioned as an uninterested transmitter of existing knowledge rather than an originator[13] and in the latter 'he did not arrive at a quantitative or analytic solution of the problem'[14] so he can scarcely be said to be of major importance.

Pascal and Fermat

Since Laplace's *Essay* (1820)[15] it has been conventional to treat the calculus of probability as beginning with a correspondence between Blaise Pascal (1623–62) and Pierre de Fermat (1601–65) in 1654.[16] In their letters they discussed some problems which apparently were posed to Pascal by the Chevalier de Méré.[17] In solving these problems they considerably advanced the application of combinatorics to the solution of probability problems, and they made the first recorded application of 'Pascal's Triangle' (which had appeared before Pascal) to the games of chance.

Although these accomplishments greatly impressed their contemporaries and successors, many modern investigators feel that Pascal and Fermat have unfairly overshadowed the earlier work of Cardano and Galileo,[18] and the later, more systematic, efforts of Huygens,[19] and James Bernoulli.[20]

Huygens

Christianus Huygens (1629–95) published the first systematic treatise on probability theory 'De Rationciniis in Ludo Aleae' ('About dice games') in 1657. This work includes the first clear exposition of the fundamental principles of the Calculus of Probability, several problems and solutions, and the first[21] discussion of the concept of mathematical expectation (the mathematical expectation of an event is the product of the probability the event will occur and the value, gain, or utility of the event to some person). This treatise 'was warmly received by contemporary mathematicians and for nearly half a century it was the unique introduction to the theory of probability.'[22]

The Bernoullis

Newcomers to the history of probability may well be excused if they confuse the various Bernoullis. Not only were there four of some importance to the subject: to add to the confusion two of them are known by three name variations each.

By far the most important is James (Jakob, Jacques) Bernoulli (1654–1705) who wrote *Ars Conjectandi* (published 1713), a general treatise on probability which also includes the famous 'Bernoulli Theorem' for predicting probable frequencies of occurrence of repetitive events.

James's brother, John (Johann, Jean) (1667–1748) was a rather unfraternal mathematical rival who did some work on probability but achieved nothing of lasting importance.

Their nephew Nicholas (Nikolaus, Nicolas) (1687–1759), did some original work in probability, notably including efforts to apply probabilistic calculations to legal problems, but is perhaps best remembered for editing the posthumous edition of James's *Ars Conjectandi.*

Finally, Daniel (1700–82), the son of John, outstripped his father in intellectual fame (especially as the originator of the Bernoulli Principle in fluid dynamics) as well as in probability theory. He is best remembered for his discussions of the Petersburg Problem[23] (what should one pay to toss a coin and receive 1 Crown if Heads appears on the first toss, 2 if Heads doesn't appear until the second, 4 if the third, 8 if the fourth, etc...?), and his development of the notion of 'moral expectation' as an attempt to solve that puzzle.[24] He was also involved in a controversy over the value of smallpox inoculation, based on the probable benefits to the individual and the community, and did some original work on the law of errors.

But his greatest theoretical contribution was the development of methods for the application of the infinitesimal calculus to probability calculations.

In this, as in most books, the name 'Bernoulli' *simpliciter* applies always to James – whom Keynes has called 'the real founder of mathematical probability'[25] and 'Bernoulli's theorem' refers to his famous law (not to Daniel Bernoulli's principle concerning fluids, as happens in some general or scientific works).[26]

Montmort and De Moivre

The mathematical history of probability theory must devote great space to Pierre-Remond de Montmort (1678–1719) and Abraham De Moivre (1667–1754) for their technical advances in the calculation

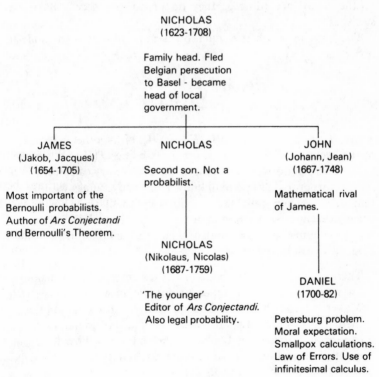

NICHOLAS
(1623-1708)

Family head. Fled
Belgian persecution
to Basel - became
head of local
government.

JAMES	NICHOLAS	JOHN
(Jakob, Jacques)		(Johann, Jean)
(1654-1705)	Second son. Not a	(1667-1748)
	probabilist.	
Most important of the		Mathematical rival
Bernoulli probabilists.		of James.
Author of *Ars Conjectandi*		
and Bernoulli's Theorem.		

NICHOLAS
(Nikolaus, Nicolas)
(1687-1759)

'The younger'
Editor of *Ars Conjectandi*.
Also legal probability.

DANIEL
(1700-82)

Petersburg problem.
Moral expectation.
Smallpox calculations.
Law of Errors. Use of
infinitesimal calculus.

Figure 1 The Bernoullis

23

of probability, but since their work is not philosophically contro-
versial we will slight them here. (Others in this category include
Thomas Simpson (1710–61), Leonhard Euler (1707–83), and Joseph
L. Lagrange (1736–1813).)

Bayes

If ever fame was achieved solely by penning a single equation, it
was done by Thomas Bayes (1702–61). Although Bayes was well
enough regarded by his contemporaries to be elected to the Royal
Society in 1742, he left few written works and little biographical
material. But in his paper, 'An Essay Towards Solving a Problem
in the Doctrine of Chances' (1763), he presented an equation (later
generalized by Laplace) which continues to divide scholars into
hostile camps according as they do or do not 'accept Bayesian
statistics.'

The form in which this equation is normally given in modern
texts[27] is:

$$P(H_k|A) = \frac{P(A|H_k) \cdot P(H_k)}{\sum_j P(A|H_j) \cdot P(H_j)}$$

where H_k is one of a series, $H_1, H_2, \ldots, H_j$, of hypothetical 'causes'
of the event A. When the a priori probabilities $P(H_j)$ are known,
this is a straightforward theorem of the probability calculus. But if
the a priori probabilities are unknown and one assumes with Bayes
that they are equiprobable (sometimes called 'Bayes's Postulate'),
one is then engaged in determining the 'inverse probability' of causes.
As an example of how controversial this method is, consider
Hogben's vitriolic remarks:[28]

> There is no conceivable factual basis for embracing this axiom
> known as Bayes' postulate; and before he answered to his
> Maker it seems that its author had not convinced himself that
> there is. However, [Richard] Price made it the kingpin of his
> exposition. Whereafter, Laplace embraced it with boyish
> enthusiasm and built on so insecure a foundation a
> superstructure of doctrine usually referred to as that of *inverse*
> probability. The adjective signifies that the doctrine licenses one

to draw conclusions about past occurrences. Among other exploits, K. Pearson, a modern disciple of Laplace, used it to prove that miracles cannot happen. None the less, one must concede the historical occurrence of at least one miracle inasmuch as many highly intelligent people have been willing to subscribe to the doctrine.[28]

We will return to this controversy in the section on the Source of Initial Probabilities in Actuarial Cases.

D'Alembert's Dissent

The most famous nay-sayer in the history of probability was Jean d'Alembert (1717?–83). It was he who opposed Daniel Bernoulli in the controversy over the value of inoculation. He also held these heterodox views (summarized by Keynes):[29]

D'Alembert has three main contentions to which in his various papers he constantly recurs:
1 That a probability very small mathematically is really zero;
2 That the probabilities of two successive throws with a die are not independent;
3 That 'mathematical expectation' is not properly measured by the product of the probability and the prize.

On all of these points d'Alembert contested against the received opinions of his time (and ours). But he may be even better remembered for yet another minority opinion of his – that the probability of at least one Head in two coin tosses is 2/3. He argued this way: Heads will either appear on the first throw (a win, in which case the second throw is unnecessary), or it will not occur until the second throw (still a win), or it will not occur at all (a loss). So there are three cases (H, TH, TT), two of which are favorable, therefore the probability is 2/3.

The opposing (traditional, correct) point of view is that these cases are not equiprobable, since the first is really shorthand for the two cases HT and HH, both of which are favorable, so that three of the four possible outcomes are favorable and the required probability is 3/4.

I presume that Keynes did not include this argument in his list because it is no longer regarded as a 'contention' but rather as a mistake. Thus Maistrov says that 'd'Alembert's name appears in the literature on probability theory mainly as an example of the fact that even certain prominent mathematicians sometimes committed errors in solving elementary probabilistic problems.'[30]

Although d'Alembert's views have not prevailed, they have found occasional advocates and partial adherents of whom the most famous early was G. L. Buffon (1707–88) and late was Emile Borel (1871–1956).

Laplace

The Classical Theory of Probability reached its zenith in the work of Pierre Simon de Laplace (1749–1827). He solved more problems and developed more important mathematical tools than any of his predecessors. He also gave the most coherent and accessible exposition of the theoretical basis of probability calculations that had ever been seen. For all these reasons Todhunter could say, 'on the whole the Theory of Probability is more indebted to him than to any other mathematician'[31] and *Collier's Encyclopedia* has ventured to describe him as 'the founder of the theory of probability.'[32]

Because of his importance and the scope of his work we will frequently take Laplace as *the* spokesman for the CTP. We will begin, for example, with his definition of probability, which has perhaps been the most widely read, quoted, and adopted explanation of probability ever given.

2 THE DEFINITION OF PROBABILITY

The theory of chance consists in reducing all the events of the same kind to a certain number of cases equally possible, that is to say, to such as we may be equally undecided about in regard to their existence, and in determining the number of cases favorable to the event whose probability is sought. The ratio of this number to that of all the cases possible is the measure of this probability, which is thus simply a fraction whose numerator is the number of favorable cases and whose denominator is the number of all the cases possible.[33]

This statement came to dominate the literature and, for many years, the theory of probability. But the definition had been in use long before it appeared in Laplace's *Essay* in 1820. It had been employed by Cardano, and made explicit by James Bernoulli:[34]

> In the game of dice for instance, the number of possible cases [or throws] is known, since there are as many throws for each individual die as it has faces; moreover all these cases are equally likely when each face of the die has the same form and the weight of the die is uniformly distributed. (There is no reason why one face should come up more readily than any other, as would happen if the faces were of different shapes or part of the die were made of heavier material than the rest.)

Thus the probability of throwing a Five is 1/6, since there are six possible cases (six ways the die can land) of which only one is favorable. But the probability of throwing an even number is 1/2 since three of the six possible cases are 'favorable' (2, 4, and 6). Calculations of this sort are essentially all that is needed to compute the odds in elementary games involving dice and cards. Completing these calculations, and more elaborate ones of the same kind, constituted much of the work of the classical theorists. In these efforts they adhered fairly closely to Laplace's 'official' definition. When they came to consider various problems in mortality and natural science, however, this definition failed them and they tended to abandon it (as we shall discuss more fully in the section on the Source of Initial Probability in Actuarial Cases). But they did *not* replace the official definition with another in terms of likelihood, relative frequency, etc. Instead they just continued to use the word 'probability' as if everyone understood its meaning, while actually employing methods and concepts which are clearly inconsistent with what they said 'probability' meant. As a result of this kind of thing Carnap says, 'it seems to me that there is no one meaning of the term "probability" which is applied with perfect consistency through-out his work by any of the classical authors.'[35] Despite this genuine ambiguity of *usage*, there is only one real *definition* of probability in the Classical writers – Laplace's definition – and that is the one we shall refer to as the Classical definition.

In addition to this *definition* of probability, however, the Classical writers include many *explanations* and 'clarifying' remarks which have aroused some controversy. These are the notorious remarks

27

by Bernoulli and Laplace (especially) which describe probability as representing a 'degree of belief.' Because of statements like these, some writers such as Kneale and Nagel have described the CTP as psychologistic or subjective.[36] While there is certainly some justification for this, I hold with Carnap[37] that what is being discussed is not the actual, contingent degree of belief, which is a matter for empirical, psychological investigation, but rather the justified or rational degree of belief, which is a matter of objective logical and mathematical fact. Seen in this favorable light, the CTP gives us rules for proceeding rationally under conditions of uncertainty; on a harsher interpretation, it is a rash and unwarranted attempt to found knowledge on ignorance and certainty on a lack of information.

3   SOURCES OF INITIAL PROBABILITY

The greatest accomplishment of the CTP is that it first made possible the quantification of probability and its mathematical manipulation by means of the probability calculus. The calculus endures to this day with little revision and is of sufficient importance to secure the reputation of those who originated it. But for our purposes the calculus is of minor interest since it is common to all major theories. In what follows, therefore, we shall not be concerned with the mathematical manipulation of probabilities, but with their origin, meaning and philosophical significance. In these matters the Classical theorists are occasionally confused, inconsistent, or just wrong. The reader should try to remember that latter-day criticisms of this sort are comparatively cheap and easy when contrasted with the extraordinary intellectual feats of the founders of probability theory. The bulk of their work is unquestioned and beyond criticism, and without it nothing that follows would have been possible.

In Dice Games

The cast of a well-made die is the very paradigm of Classical probability. It is simple, important to gamblers, easy to understand, and, one might think, obvious. Indeed historians of probability have sometimes seemed less concerned with how this basic problem was

28

solved than with the puzzle of why it was not solved before. As noted in our historical sketch, gambling has been with us at least since classical antiquity. Yet nowhere in literature does anyone calculate the probability of throwing a Five before the fifteenth or sixteenth century! How can we account for this extreme lag between practice and theory?

David suggests the explanation is that 'the philosophic development' of the time was unsuited to 'the construction of theoretical hypotheses from empirical data.'[38] Thus Cardano's calculations, like Galileo's equations of motion, had to wait on the development of a *zeitgeist* which accepted and even encouraged quantification and scientific explanation.

A dissenting view comes from the Marxist historian L. E. Maistrov who says it is a 'widespread false premise that probability theory owes its birth and early development to gambling.' Instead, he emphasizes the connection of probability with 'other sciences and problems' and especially with 'the rise of capitalism, when commerce and monetary transactions, particularly those connected with actuarial operations, were developing and when various new institutions were established.'[39]

If we leave this puzzle to the historians, we can at least say that much of the early work in probability was concerned with games of chance, and especially with dice (playing cards are a later historical development, as well as a more complicated problem). In this early work the basic assumption is that each face of the die is equally likely to turn up. Since there are six such sides, one of which must turn up, the initial probability of each is 1/6. From this initial or a priori probability follow most of the early theorems and calculations in probability theory.

It may be that the equiprobability of the different faces was first conceived as an empirical hypothesis.[40] But it was not long before James Bernoulli developed a rule which was supposed to give a method for finding the initial probability in many cases (including dice games) and which came to be considered a fundamental part of the CTP. This rule was originally called the Principle of Non-Sufficient Reason but we shall follow Keynes's suggestion and refer to it as the Principle of Indifference.[41] The fundamental idea is that *alternatives are always to be judged equiprobable if we have no reason to expect or prefer one over another.* This principle is the object of most of the criticism (and even scorn) which has been directed at the CTP. Its unrestricted use seems to imply, for example, that any proposition of whose truth or falsity we are ignorant is exactly as probable as its negation. Still, the Principle of Indifference is the major theoretical justifications for the equiprobability of alternatives in dice games and is therefore the principal source of initial probability in such cases.

Now it is an unfortunate fact of life that not all games are fair and not all dice are unbiased. What, then shall we say of biased dice, where ordinary language would say that the faces are *not* equally probable? Laplace discusses such problems in his *Essay* with the following results.

First of all, if we know the die is biased but do not know *how* it is biased, we still have no reason to prefer one face over another and, by the Principle of Indifference, they remain equally probable.[42]

Second, if we know how and to what extent the die is biased, we can use the probability calculus to compute the odds. In the case of a biased coin, for example, Laplace says[43]

> In order to submit this matter to calculus let us suppose that this inequality increases by a twentieth the probability of the simple event which it favors. If this event is heads, its probability will be 1/2 plus 1/20 or 11/20, and the probability of throwing it twice in succession will be the square of 11/20 or 121/400.

He goes on to consider various ways of computing probabilities resulting from bias and various devices for reducing the effects of the bias. What he does not do is to explain the origin, meaning, or justification of the value attributed to the bias. Once this is known, all else follows by the simple rules of the calculus – the crux of the problem is establishing the original probabilities. We might do this by constructing a die to have a certain probability, or we might do it by using some sort of inverse method (to be discussed in the next section) in order to derive probability from observational data. The important point is that neither of these methods is compatible with the official structure of the CTP because

1 The probabilities are not derived from the Principle of Indifference, which is often described as the only valid source of initial probabilities in the CTP.

2 Furthermore, there are no equally likely cases here, so there

can be no ratio of favorable to all possible cases – but this is the only official definition of probability.

In Actuarial Cases

We have seen that the CTP breaks down when biased dice are introduced to the game. In actuarial cases, the official definition of probability is almost completely ignored, out-stripped, confounded – yet Classical theorists did much to lay the foundations of statistics and sound actuarial practice. Here more than anywhere it becomes obvious that the concept of probability actually employed by Classical theorists is far broader than the official definition laid down by Laplace. It might seem appropriate in such cases to develop the recurring concept of 'degree of (rational) belief' into a theory which is explicitly logical or explicitly subjective. On the other hand, it is quite clear that Laplace and Daniel Bernoulli (for example) were investigating mortality tables and the incidence of smallpox in an attempt to discover some objective facts about the world – facts which might very well be described as relative frequencies of occurrence. So the Relative Frequency view of probability is also foreshadowed in this early work!

All of this should remind us that Bernoulli and Laplace and their ilk were not 'adherents of the Classical Theory of Probability' whose work was an attempt to apply that theory to reality; rather, they were pioneering investigators of the concept of probability who developed one part of that concept thoroughly enough for us to call it a theory but who also worked on other aspects of the problems. When the 'equiprobable alternatives' method was practical, they preferred to use it, but they freely abandoned it whenever it was inapplicable.

The simplest and most widespread alternative to the Principle of Indifference was the use of actuarial tables as a source of initial probabilities. The first such useful tables were perhaps the plague mortality tabulations prepared by John Graunt of England during the Black Death. The Classical theorists used such tables as a source of information about what had happened in the past. The values thus obtained were then employed as probabilities for future occurrences. The simplest justification for this is James Bernoulli's assumption that things in the future will exhibit the same pattern as things in the past.[44] A more sophisticated idea is that statistical evidence reveals the underlying probability of occurrence for certain types of events, since we know that the actual outcomes must be proportional to the probability. This is Daniel Bernoulli's reasoning in his discussions of marriage and mortality. Finally Laplace, building upon the work of Bayes, developed the notion that observed results give us evidence which can be used to compute the probability that the underlying probability has a given value within its possible range.[45] In this form the use of mortality tables can almost be forced back into the mold of the Principle of Indifference by the following reasoning.

Consider a mortality table for a certain population. An individual taken at random might be represented by any one of the entries in the table. The Principle of Indifference therefore tells us that it is *equally* probable that he will be represented by any given one or the other (since we have no reason to 'expect' or 'prefer' one over the others). Given this initially equal probability distribution, it follows that the individual's probability of death is directly proportional to the death-entries in a given year.

It would seem, then, that the CTP *can* handle actuarial cases, *if it is provided with a valid statistical description.* So the problem of initial probability in these cases is transformed into the problem of constructing valid actuarial tables.

Now Laplace was convinced that if we had mortality tables of infinite extent we would have a perfect value for the probability of death (since all possible cases would certainly be included).[46]

As our actual tables are obviously finite, however, is it possible to assess their accuracy, reliability, or similarity to the 'true' table? This is where Bayes's Theorem comes in.

Suppose we were concerned with only 4 people and we knew that 3 had died at certain ages while the fourth was still alive. Suppose further that there are only 2 mortality tables which can possibly give the true probabilities of death for these people, and that table X is a priori twice as likely to be true as table Y. In this case we can use our a posteriori knowledge of the deaths, together with our a priori knowledge of the probability of each table and our knowledge of the probability that just this pattern of deaths would occur if X or Y were true to calculate the probability that a given table is the correct one.

If we use our earlier statement of Bayes's Theorem

$$P(H_k|A) = \frac{P(A|H_k) \cdot P(H_k)}{\sum_j P(A|H_j) \cdot P(H_j)}$$

we have

$$\begin{array}{l} P(\text{mortality table} \\ X \text{ is true cause} \\ \text{given these} \\ \text{deaths}) \end{array} = \frac{\begin{array}{l} P(\text{this sequence of} \\ \text{deaths if } X \text{ is true}) \end{array} \cdot \begin{array}{l} P(X \text{ is true} \\ \text{a priori}) \end{array}}{\left[\begin{array}{l} P(\text{this sequence} \\ \text{of deaths if} \\ Y \text{ is true}) \end{array} \begin{array}{l} P(Y \text{ is true} \\ \text{a priori}) \end{array}\right] + \begin{array}{l} (\text{the} \\ \text{numerator} \\ \text{above}) \end{array}}$$

By hypothesis all the quantities on the right are already known, so it is easy to compute the probability that table X is the true description. The probability for Y could be computed likewise, but it is obviously equal to $1 - P$ (X is true given these deaths). In such an ideal situation, Bayes and Laplace can uncontroversially tell us the probability that a given mortality table is the true one.

But again we are slipping away from the official definition of probability. It is true that 'the probability of this sequence of deaths if X is true' can be accommodated by the definition when each person is construed as equally likely to correspond to each entry. But consider the a priori probability of each table – except for the singular case where they are equiprobable, we can find no 'equally likely cases' on which to base such probability. We are dealing here with a pure, undefined sense in which we understand what it means to say 'the probability of Y is 1/3 and of X is 2/3,' but according to the official definition we *should not* understand this.

Nevertheless, we resolved earlier not to fret excessively if our classical theorists strayed too far from the theory we attribute to them. Besides, these same a priori probabilities have been the source of enough fretting from another direction to far overshadow our definitional quibbling. You see, what Bayes and Laplace did in less-than-ideal cases was to assume that all possible causes *are* a priori equiprobable. Such an assumption does guarantee equally likely cases (for what it's worth), but it also seems to many thinkers to be completely and totally unjustified. The essence of the quarrel about 'Bayesian statistics' is not about Bayes's Theorem at all, but about Bayes's Postulate which asserts this equiprobability of causes. Without entering too deeply into the controversy we can just note that critics argue that the postulate is mathematically and philosophically unjustified and can be shown to lead to wildly incorrect 'solutions' in many cases. But, the Bayesians rejoin, in the first place such wild errors are the exception rather than the rule; in the second place the magnitude of any error due to the postulate decreases rapidly as data accumulate, and in the third place, what else can we do in a less than ideal situation where we need to know such a probability but the a priori probabilities are unknown – we must use the Postulate or just give up.

So in controversial cases Bayes's Theorem does not let us 'go from frequencies to probabilities' but instead proceeds 'from frequencies plus assumptions about a priori probability' to its conclusion. (The Postulate, by the way, closely resembles the Principle of Indifference in telling us to treat things as equally probable if we have no reason not to. It is therefore subject to the same kind of criticism and disparagement as is that Principle.) Furthermore, the conclusion or output of Bayes's Theorem is not 'probabilities' *simpliciter*. Even in ideal cases the best we can get is 'the probability that the probability of *E* is *x*.' Thus even a Bayesian should be careful about proclaiming the old saw that 'Bayes's Theorem lets us go from frequencies to probability,' but it remains true that Bayes's Theorem is one of the most widespread and powerful methods for deriving initial probabilities in actuarial cases within the general spirit of the Classical Theory of Probability.

4   PROBABILITY OF SINGLE EVENTS

The Classical theory, unlike the Relative Frequency view we will examine later, finds no difficulty at all in the notion of the probability of a single event. Since we are to compare the number of ways that an event *can* succeed (not 'has succeeded,' or 'will succeed in the long run') to the total number of ways it can occur (ditto), we do not require consideration of a series of events in order to develop a probability value and thus we have no problem (as the RF theory does) in returning from that series to apply the value to a single case. Not only is it possible to develop a Classical Probability value without appealing to empirical evidence, some critics maintain that it is necessary that we do so because the theory has no mechanisms for proceeding otherwise and is, in fact, incapable of learning from experience (more on this later).

That Smith will roll a Five

We have said that this can well be taken as a paradigm case for the Classical theory. As Smith prepares to throw the die he knows and all of us around the table know that there are six faces on the die which may 'come up,' and that only one of these is a Five (presuming we have checked to ensure that the die is properly imprinted). The Principle of Indifference tells us that if we have no reason to expect or prefer one of these outcomes, we are to treat them as equally probable. As there are six such alternatives, the probability of each is 1/6.

The power and simplicity of this reasoning has made this method almost universal among gamblers and common among laymen. An astonishing number of extremely complex problems in probability theory have been solved, and usefully so, by calculations based entirely on the assumption of equiprobable alternatives. This remarkable success story may fade from the reader's mind as we proceed to develop the inadequacies, confusions, and contradictions involved in the CTP – try to remember that our strongest ordinary conception of probability and most of our mathematical successes are based on just this picture of Smith standing there with an equal chance of throwing a Five.

But however successful the CTP has been in dealing with normal dice games, in this rotten world you can be sure that not all dice are 'normal' or 'fair.' What are we to do about those cases?

Suppose, for example, that as Smith prepares to roll the die Jones comes by and whispers in your ear, 'It's loaded.' Don't delay, place your bet to exploit this knowledge and get rich quick!

But how? What shall we bet on? Does Jones mean Smith has slipped in a loaded die and we should bet with him on the Five? Or does he mean that the house is crooked and we should always bet against the shooter? Or has Jones himself loaded the die intending to make a side bet on the Three? What are we to do?

As we saw above, the bare knowledge that the die is a loaded has no effect at all on the probability that Smith will throw a Five!

Unless we know which number is weighted we still have no reason to prefer one over the other and must continue to regard them as equally probable. (It is clear that in some sense we now know that they are not 'really' equiprobable – this problem will be discussed in the section on Absolute Probability.)

So for the innocent Smith the probability of a Five is 1/6, and since our guilty knowledge is insufficient it remains the same for us. But what about Jones? Suppose he knows how the die is loaded; what can the CTP do for him? If he knows the extent of the influence of the loading, he can plug it into the calculus and generate a probability as Laplace did in the quotation above.[47] If he does not already know the extent of the influence of the loading but only its direction, there is no formal way for the CTP to compute the probability. Indeed there is no way even to describe it in the official theory, since alternatives are no longer equally probable. Here again the Classical theorists' understanding and use of the concept of probability outstrips our official definition. But here Jones cannot use Bayes's Theorem to help him out, since we are talking about a single throw of the die, and he has not the time to collect evidence.

Still, the CTP can advise Jones on what to do. Since the time of Huygens the Classical theorists (and almost everyone else) have employed the concept of the mathematical expectation of an event and have assumed it to be a rule of rationality that one should maximize that expectation. If each number pays off equally, the difference in the expected value of each is directly and only proportional to that number's probability of occurrence. Since Jones knows which number has the highest probability of occurrence, he *ipso facto* knows which one has the highest expected value and, to be rational, should bet on that number.

Detesters of triviality may feel it no credit to a theory that it uses big words and complicated mathematics to tell someone to bet on the loaded number. A little reflection should show, however, that there are many cases in which one should *not* bet on the loaded number (if the payoffs are disproportionate, for instance, or some outcomes are naturally more probable than even a loaded number, or if combinations and negative bets are more lucrative, etc.). As the situation increases in complexity, even keen common sense and natural ability will soon feel the need for assistance from the mathematical structure developed for Smith's improbable Five.

Detractors of the Classical theory in general and the Principle of Indifference in particular have often argued that the probability in a dice game is not based on our ignorance of any distinguishing features of the die faces, but rather on our knowledge, based on collective experience, that Five comes up 1/6 of the time. This point is partially rebutted by our discussion of a die that is known to be biased but not in any particular direction. Here we can be sure that Five will not come up with the normally observed relative frequency – it appears either more or less frequently than one sixth of the time. And yet the fair bet, in this state of ignorance before the first roll, is based on a probability of 1/6 (5 to 1 odds).

Perhaps the critics will respond that past experience has shown that each side of a die gets loaded with approximately the same frequency as the others, so that again our judgment is based on experience rather than ignorance. But whatever reasonableness this argument might have when applied to dice games vanishes when we consider a unique, single election.

If Smith, Jones and Robinson are running for mayor, it is quite clear that the CTP can give us a precise numerical value for the probability that Smith will be elected – but only if we know nothing at all about the election and the candidates. Once we acquire knowledge of the age, race, sex, occupation, party, etc. of the candidates, we become less able to assign definite values to the probability (though it might be quite clear which values increase and which decrease from the original 1/3).

The reason, of course, is that the Principle of Indifference allows (requires) us to treat the alternatives as equally probable so long as we have no evidence to the contrary. Using this ignorance as our justification, we should accept any bet offered by equally ignorant persons (or devices, mechanisms, situations) at odds better than 2 to 1. The amazing thing, which so impressed Quetelet and Poisson, is that if we act on such ignorance repeatedly we can be almost certain of being right one-third of the time (Bernoulli's theorem) while this certainty is not readily available to any other method (some selection procedures give sub-random success in some situations – as if you always pick the Socialist Workers' Party candidate in Mississippi elections).

This 'knowledge from ignorance' has been the most controversial claim of the CTP. We shall examine both sides in later sections of this chapter – for now, remember that Smith's probability of being elected mayor is an excellent example of it.

5 PROBABILITY OF REPETITIVE KINDS OF EVENTS

The Classical theory makes no use of series and repetitive events in its definition of probability and in its clearest applications, but CTP theorists have been interested in repetitive and actuarial processes since the very early stages of theorizing. The most famous tools which they developed for dealing with such problems are Bernoulli's Theorem, Bayes's Rule, and Laplace's Rule of Succession.

That a Dice Throw will be Five

We saw above that the Principle of Indifference gives a probability of 1/6 that a single throw will be a Five. Now let us consider a series of 100 such throws – how many can we expect to be Fives? Common sense (and fairly simple calculation) tells us that the most likely number is 17 (1/6 × 100 = 16.666) and that we should 'expect' it to be 'fairly close' to that value. In particular, the probability that the number of Fives will be between 14 and 20 inclusive (17 ± 3) will be approximately 0.65. Let us call the number of Fives in a series of N throws S (for success, a common usage). Then, in our example, $N = 100$ and we computed the probability that $S = 17 \pm 3$. The number of successes, S, divided by the number of trials, N, gives the frequency of success (strictly, the relative frequency of success, since S is called the absolute frequency of success) and we shall call this F. Thus we were calculating also the probability that $F = 0.17 \pm 0.03$. Now suppose that we increase the sample size to 1,000. The same frequency of success F will be represented by a number of successes S, between 140 and 200. But now the probability of S falling in this range is 0.99.

Bernoulli's Limit Theorem says that continuing to increase the sample size N will cause the probability that F will fall within the desired range to continue to approach 1, and that it can be made to approach 1 as closely as we wish for any desired range of success no matter how narrow.[48]

In symbols,

$$\text{Lim}_{N \to \infty} P(|Np - F| < E) = 1$$

where F is the relative frequency of success, p is the probability of success in each individual trial (which must be independent, by the way), E is any fixed degree of variation we choose, and N is the number of trials.

The Classical theorists and their successors have claimed that this theorem demonstrates that simple probabilities will be realized in experience as relative frequencies. Thus the probability 1/6 of a Five on a single throw guarantees that in a sufficiently long series of throws the relative frequency of Fives will be very close to 1/6.

That is how the CTP deals with series of events. It is not an uncontroversial method and we shall present several criticisms of it in the appropriate section, but it has led to considerable practical success and is certainly not to be dismissed out of hand.

That a Thirty-year-old will get Married

When we ask the probability that a bachelor will marry this year, it seems there are only two alternatives: he will or he won't. If we knew nothing further about the matter and if we were very liberal in our use of the Principle of Indifference it might be that we would treat these alternatives as equiprobable and call the probability 1/2.

But of course we know that the general probability of marriage is not as high as 1/2 and only a few extremists will apply the Principle of Indifference to a statement and its negation. So what can we do here?

As we indicated above,[49] what the Classical theorists did was to use Bayes's Rule in some cases and to consider actuarial tables as direct sources of probability in other cases.[50] A complete theoretical justification on classical lines can be provided for these methods only in the extremely unlikely case where we know an actuarial table is complete and accurate or where we know all the possible alternative tables and the a priori probability that each is true. Despite this gap between theory and practice, these methods were effective enough to guarantee profits to the early insurance companies and respectability to the early statisticians.

But there is another method for dealing with successive events which has gained even more notoriety than Bayes's Rule or the Principle of Indifference – that is Laplace's Rule of Succession. As stated in the *Essay*: 'Thus we find that an event having occurred successively any number of times, the probability that it will happen again the next time is equal to this number increased by unity divided by the same number, increased by two units.'[51]

After *N* successes in a row, then, the probability of success on the next trial is

$$P = \frac{N+1}{N+2}.$$

Laplace derives this rule from a version of Bayes's Rule plus some assumptions. The import of the assumptions is that there must be an infinite (at least very large) number of possible constitutions of the universe of discourse with each one being equally probable. The rule is sound, for example, when we pick one urn out of a series exhibiting every possible combination of red and white balls, then draw *N* balls (with replacement), all of which happen to be red. The probability[52] that the next ball will be red is then

$$\frac{N+1}{N+2}.$$

The application which Laplace himself made of the rule was to calculate the probability of tomorrow's sunrise: 'Placing the most ancient epoch of history at five thousand years ago or at 1,826,213 days, and the sun having risen constantly in the interval at each revolution of twenty-four hours, it is a bet of 1,826,214 to one that it will rise again tomorrow.'[53]

This example has been criticized for its historical data (why 5,000 years? Aren't we surer of tomorrow's sunrise than of one that long ago?) and its metaphysical assumptions (Is the universe really like an urn with all compositions equally possible?). The general application of the rule has been further criticized on the grounds that it leads to absurdities (If the first person I meet today is red-haired and deaf the probability that the second person I meet will be red-haired and deaf is 2/3) and contradictions (If I have met 2 red cars, 3 white cars, and 4 black cars, the probability that the next will be red is 3/11, white 4/11 and black 5/11, for a total probability of 12/11).

6 ABSOLUTE PROBABILITY

In the Classical theory there is no clear notion corresponding to our idea of absolute probability. It is evident that probability can be relative, as when one person's knowledge of his poker hand gives him a probability for drawing an ace different from that computed by his neighbor. In such cases probability is relative to our knowledge – in cases using the Principle of Indifference one might say that probability is relative to our ignorance. In either case, probability is apparently not empirical, it has to do with our beliefs (the psychological interpretation) or rational methods of inference (the logical interpretation) but not with any empirical property of things.

On the whole the logical interpretation seems to be the most justified, since it can account for the talk about degrees of belief and the precise rules which are supposed to govern everyone's probability conclusions.[54] On this view, the most reasonable interpretation for the absolute probability of event E would be 'the probability we would assign to E if we possessed all relevant evidence and all necessary mental abilities.' But Laplace's demon and several remarks of Bernoulli's[55] show that this limiting case is one of certainty: in such an ideal situation we would *know* if E would happen or not. In accordance with this interpretation of the CTP, therefore, the absolute probability of every meaningful proposition must be either 1 or 0.

It should be noted that many passages in Classical writings do suggest an empirical rather than a logical definition. Most notable, perhaps, is that the official definition refers to the number of ways in which an event *can* happen, rather than the number of ways we *think* (or should think) it can happen. Also, Classical writers frequently speak of 'unknown' probability, which should not exist in a logical or subjectivist interpretation. Most frequent of these is the unknown composition of an urn; true actuarial tables are also said to be unknown and just 'approached' by existing tables. Considerations of this sort just might move one to think that the CTP includes a notion of physical or empirical probability. Even if this is the case, however, the thoroughgoing determinism which we shall presently discuss requires that for any individual event it either will certainly happen or will certainly not happen. Even on this interpretation, then, the absolute probability of E must be either 1 or 0.

7 PHYSICAL CHANCE

The Classical theorists were determinists. They wrote, for the most part, in the grip of Newton's clock-like cosmology which so dominated European thought in that period. They were convinced that the events in nature were links in a causal chain, so that each one is determined by those which precede it and, in its turn, helps provide 'Sufficient Reason' for events to follow.

This determinism occurs clearly and explicitly as early as Bernoulli ('everything in the world occurs for definite reasons and in definite conformity with law...')[56] and Montmort ('nothing depends on chance... all things are regulated according to certain laws...').[57] But there is one statement of the determinist thesis which is so compelling in its imagery and so powerful a summation of determinist thought that it can nearly stand alone as an expression of that metaphysical view. I am speaking, of course, of Laplace's demon:[58]

> We ought then to regard the present state of the universe as the effect of its anterior state and as the cause of the one which is to follow. Given for one instant an intelligence which could comprehend all the forces by which nature is animated and the respective situation of the beings who compose it – an intelligence sufficiently vast to submit these data to analysis – it would embrace in the same formula the movements of the greatest bodies of the universe and those of the lightest atom; for it, nothing would be uncertain and the future, as the past, would be present to its eyes.

Stated in simpler materialistic terms: If the Demon knew the position and momentum (etc.) of every particle in the universe at a given time, he could in principle predict or retrodict every event in the history of the universe. This is the type of determinism strongly suggested by Newton's Laws of Motion, which successfully described and predicted the motion of the planets and claimed also to deal with microcosmic events. Some think this view should perish with the physics that inspired it, so that indeterminism and Free Will are swept into office along with Relativity Theory and Quantum Mechanics.[59] But at the time of Laplace, Newtonian cosmology ruled Science and determinism was *au courant*.

In a deterministic world a chance event cannot be one which 'might happen one way or the other,' since all events *will* happen in an exactly specified manner. Chance, then, must be a psychological or epistemological, rather than a physical, phenomenon. This is the official position of the CTP, that when we talk about chance we are referring to *our* inability to predict events, not to any genuine randomness or indeterminacy in the events themselves. This, I think, is as clear and definite a doctrine as one can find in the CTP.

Of much less importance, but of some interest, are later attempts to specify within this deterministic framework some type of events which are peculiarly 'chance' events and are especially suitable for treatment by the calculus of probability. Poincaré, for example, emphasizes events which have a multitude of causes or which are such that slight differences in initial conditions lead to great differences in the eventual outcome (the toss of a coin, for example, or the roll of a die). Cournot, on the other hand, finds the essence of chance in the convergence of independent causal chains, so that there may be a perfectly good causal explanation for a worker's dropping a brick from the top of a building and there may be an equally good cause for my walking along a certain sidewalk on a given afternoon, but if these causal chains come together in such a way that the brick 'happens' to land on my head, we are much less ready to give an 'explanation' of this event and are inclined to think it was due to 'chance' or 'bad luck'. As Keynes notes, however, even such attempts to define chance took place in a deterministic context, so that there really was no such thing as pure physical chance in the universe of the CTP.

We have, then, a world in which everything is predictable – in principle. But when we live out our lives we see that predictability in principle cuts no wood and draws no water: what we need is predictability in practice. Perhaps Laplace's demon could tell us if our cow will get sick or not, but *we* can't tell. We are surrounded by events whose causes are obscure, unknown, or incalculable; it is beyond our ability to predict them. But even though predictability in principle vanishes when we turn to individual events, it reappears when we consider mass or repetitive events! If Laplace cannot tell us whether one cow will get sick or not, he *can* tell us how many sick cows to expect in a given year. And if the herd is large enough and conditions are relatively stable, we can use his prediction to make money on the farm.

This predictability *en masse* is the source of the vast profits of insurance companies. It was also the source of the rather excessive enthusiasm for probability theory conceived by Poisson and Quetelet, who found in the Law of Large Numbers a most remarkable uniformity extending to all human endeavors and all parts of nature. In a later section we will detail some criticisms of the CTP's treatment of the Law of Large Numbers. Yet it cannot be denied that a considerable practical advance occurred when the Classical theorists pointed out Nature's tendency to operate with stable frequencies and normal distributions. It is interesting to note that the kind of predictability thereby introduced would be exactly the same even if the world really were ruled by 'random chance' rather than rigid determinism. It is therefore perhaps not quite as strong an argument for lawlike uniformity in the world as Poisson and Quetelet thought. Still, it is also in accord with the official position of the CTP, that all events are determined and there is no such thing as physical chance.

8 THE METAPHYSICAL STATUS OF P

For the most part, Classical theorists held that probability is not a genuine part of metaphysical reality but is a human invention cleverly designed to assist us in making rational choices when we have less than complete information. This interpretation especially accords with their frequent remarks that probability is a measure of our ignorance, or a product of it. It likewise fits with the uniform determinism which is assumed to govern the universe. In such a Newtonian-clockwork universe, there is no such thing as probability – it has no metaphysical reality at all.

I think these reasons are sufficient for us to conclude that there is no constituent part of the universe which can be called 'probability' in the sense of 'going one way or the other,' 'depending on chance,' etc. Yet there remain two ways in which real features of the universe are involved in probability: (1) in cases like urn and dice problems, the constitution of the urn and the number of faces on the die determine the probabilities – and these are physical (hence, metaphysical) realities; (2) events like infant mortality do occur in nearly fixed frequencies over long stretches of time, so that we can and do base probabilities on real properties of collections of real events.

In the first of these cases, let us consider the urn problem. Suppose there is an urn before us which we know is either urn *A*, with 2 red and 2 white balls, or urn *B*, with 4 white balls, Our probability of drawing a red ball is then 1/4.

$$P(A) = 1/2 = P(B)$$

$$P(R/A) = 1/2$$

$$P(R/B) = 0$$

$$P(R) = [P(R/A) \times P(A)] + [P(R/B) \times P(B)]$$

$$= [1/2 \times 1/2] + [0 \times 1/2]$$

$$= 1/4 + 0$$

$$= 1/4$$

But, if Jones knows that we are facing urn *A*, he knows that our probability of drawing a red is 'really' 1/2. This is what Laplace often calls an 'unknown' probability, and it seems to be a feature, not of our knowledge, but of the urn's true composition. It seems to be a metaphysically 'real' probability, while the earlier one seems only relative to our ignorance.

This, however, is only an illusion, based on the fact that Jones has better information than we do and therefore seems closer to reality. Yet his probability is just as much based on ignorance as our own. He knows that there are 2 red and 2 white balls in the urn but he doesn't know which we will pick, so he assesses the probability at 1/2.

Now suppose another person, Smith, comes into the room, and *Smith knows we will pick a red*. Then the probability for him is 1, and Jones's probability no longer seems like the real thing.

Smith, of course, is Laplace's demon, capable of knowing all future facts about the universe – and there are such facts to be known! We *must*, therefore, be deceived when we think there are 'real' probabilities other than the degenerate ones of 1 and 0. But wherein lies the deception?

I think the answer is that the probabilities which we normally think of as 'real' are those which are predicated on the best knowledge normally available to human beings. We normally know how many faces a die has and whether or not it's loaded, but we *don't* normally know precisely how it will be thrown, what air currents will affect it, etc. And we certainly do not normally know how it will land. Therefore the best state of knowledge we can normally achieve tells us that the probability of a Five is 1/6. A half-smart demon, or a very good dice manipulator, might know that this time the probability is 'really' O – it all depends on your state of knowledge. Since the paradigm case for human knowledge of an urn problem involves complete knowledge of the composition of the urn but *not* complete knowledge of which ball will be chosen, it is perfectly natural that the paradigm probability (for us) is that probability which can be computed using that paradigm knowledge. Therefore the composition of the urn determines a probability value which is special to us, but not necessarily special to the universe. (In fact, if the Classical theorists are right in their picture of a deterministic universe, our 'real' probabilities, determined by the composition of the urns, are necessarily false if they are not 1 or 0.)

Turning now to our second problem, probabilities are sometimes thought to be real because they are based on real frequencies. Let us take it as given that there are fixed frequencies of death and the like which have been established in the past and which we know, by some 'valid' form of induction, can be counted on to continue in the near future. Does this establish that there are metaphysically real probabilities?

Well, first of all, it establishes that there are metaphysically real relative frequencies, even real frequencies of future occurrences (this, incidentally, agrees well with the CTP's deterministic metaphysics). To a frequency theorist, that is sufficient to establish that there are real probabilities. But to the CTP, probability is not a frequency (despite occasional confusions and unclarity in the writings, we take this to be the official position). Instead, probabilities are based on the existence of computable equiprobable alternatives.

I have suggested above that the use of mortality tables is theoretically justified in the CTP by the fact that we are equally likely to be represented by a favorable as by an unfavorable entry, so that the ratio of favorable entries to all entries constitutes our probability of survival. Now let us suppose that the death frequency of white males of my age group in Florida is 0.02 per year, and that this frequency is stable for groups as small as 1,000. It is then possible to construct a mortality table containing 20 unfavorable entries and 980 favorable entries, knowing that I am represented by one of the entries in this 'real' mortality table but not knowing which one. We can then know that 20 of this group of real people will die, and the Principle of Indifference tells us that there is a 0.02 probability that

I will be one of them. This is, *ex hypothesi*, a probability based on reality – but is it a 'real' probability?

According to the CTP, it is the probability of my death *if* I have no reason to 'prefer' one or the other alternative (not in the sense of a death wish but in the sense of a rational expectation). Now it happens that I know that I drive a sports car and I know that sports car drivers are slightly more likely to die than Cadillac drivers or pedestrians. Therefore my 'real' probability of death, based on this knowledge, is slightly greater than 0.02. On the other hand, I am healthy, educated, very well fed, and receive good (socialized) medical care. Therefore my 'real' probability of death is somewhat less than before. In principle, each of these complications produces a changed but calculable probability of death (assuming each is associated with a stable frequency). But which is my 'real' probability of death? We normally think that the 'best' probability is that which is based on the 'most' knowledge. But Bernoulli and Laplace insist that in the pursuit of knowledge the limiting case is – certainty. Again, Laplace's demon *knows* whether I'm to die or not and there's an end to speculation. In their universe, I either will or will not die, and there's no such thing as a 'real' probability of my death.

If there are stable frequencies in nature, as we assumed, that fact allows us to predict future mass events with near certainty. It also gives us grounds for computing probabilities which are valuable (some more so than others). But even though the facts are real, the probabilities are always relative to our knowledge (which cases are equiprobable to us?). Therefore, metaphysically real frequencies do not determine metaphysically real probabilities, mortality tables and 'unknown probabilities' notwithstanding.

9   THE EPISTEMOLOGICAL STATUS OF P

It would be stretching the matter only a little to say that, for the classical theorists, probability *is* an epistemological phenomenon. That is to say, probability is not some feature of the world, which we seek to discover, but rather it is a way of dealing with the world, based solely on our own knowledge and mental faculties, and helping us to project the future out of our descriptions of the present. It is, according to Bernoulli and Laplace, a substitute for exact knowledge – since knowledge is the very subject of epistemology, presumably its substitute is likewise epistemological.

The curiously unepistemological feature of Classical probability is that it notoriously bases its projections on ignorance rather than knowledge. This is especially true of the Principle of Indifference, which has been said to require 'an equal distribution of ignorance' in order to assign numerical values to probabilities. Remember the mayor's race we described above, where we knew the probability of Smith's winning only if we knew nothing at all about the race – when we learned something significant, we lost our ability to give exact odds. This is the kind of 'knowledge out of ignorance' situation which has attracted many thinkers and repelled many more. It seems to denigrate careful investigation and collection of information and rely instead on mathematical mumbo-jumbo to produce as if by magic the exact probability of occurrence of something we may never have experienced.

As one might expect, the critics have somewhat overstated the case here. It is not just our ignorance that is important, Laplace says, 'Probability is relative, in part to this ignorance, in part to our knowledge.'[60] In order to deal with the (somewhat uncharacteristic) probabilities of death, for example, it is clear that a statistical investigation must first create a mortality table. In the simpler case of a dice game, our first task is to identify the equiprobable alternatives. But in acquiring this little bit of knowledge the CTP encounters a major epistemological problem: How do we justify the belief that all faces of the dice are equally probable to turn up?

This question is of fundamental theoretical importance because, as I shall argue throughout the book, there is little controversy about how to deal with established probabilities. All theories share in common the calculus of probability, they differ primarily on (1) the definition of 'probability,' and (2) the method of establishing initial probabilities.

Now if we *know* the faces of the die are equally likely to come up, and if we know that one and only one must come up, it is mathematically necessary that the probability of each be exactly 1/6. But how do we come by that first bit of knowledge?

Relative frequentists and other critics of the CTP argue that we learn the equiprobability of the faces the same way we learn other empirical facts – through experience. Generations of gamblers have carefully noted the fall of the dice, and generations of dice makers have labored to insure that the gamblers have seen each face equally often. That, claim the critics, is how we know the faces are equally probable.

Not so, say the Classical theorists, the faces are equally probable because Bernoulli's Principle of Indifference (Principle of Non-Sufficient Reason) tells us to treat equipossible alternatives as equally probable if we have no reason to treat them otherwise.

But now our epistemological problem has become two problems: how can we identify equipossible alternatives, and how can we justify the Principle of Indifference itself? The first of these questions will be prominent in the section devoted to criticism of the CTP, so I will now pass it by to deal with the second.

The Principle of Indifference has never been argued *for* to nearly the extent it has been argued *against*. The Classical theorists themselves seem hardly to have felt a justification necessary. They were concerned with developing and expounding a system which worked. Philosophical analysis, criticism, and defence naturally came later. The most notable defenders of the Principle among later thinkers were Keynes and Carnap. Each of them decried the unrestricted and uncritical use of the Principle which had led to excesses and contradictions in the past and each modified, restricted, and generally prettied up the Principle before incorporating it in his own system. In the end, Keynes relied on our direct intuition to show us the truth of the Principle, while Carnap appealed to its simplicity and the fact that it worked well in practice. If Laplace had felt called upon to justify the Principle, I'm sure he would have approved of all three of these considerations, but I really think that he thought it was just obvious. After all, what *other* value could you reasonably assign to the probability of rolling a Five if you had no reason to expect one face or the other? It just *has* to be 1/6.

For my own part, I think the Principle becomes more plausible when it is clearly removed from the idea of the probability of occurrence of an event and explicated instead in terms of the notion of a random guess. I will explore this idea further after I present the main criticisms of the Principle in the next to the last section of this chapter.

10   RATIONALITY OF PROBABILITY BEHAVIOUR

In addition to developing the calculus of probabilities, the Classical theorists also formulated two other ideas which are of major importance to the general study of rationality: the concept of Mathematical Expectation, and the Principle of the Diminishing Marginal Utility of Wealth.

The first of these we have attributed to Huygens (despite Keynes's pressing of the Leibniz claim, see note 21 to chapter II). It has become so well known that most readers will be familiar with it already, but it is always well to review things and ensure that we are talking about the same concept.

The Mathematical Expectation of an event is the product of the probability of the event and its value (sometimes, utility) to an individual.

$$ME(A) = P(A) \times V(A)$$

This, of course, is nothing more than a definition. The Principle of Rationality which makes it important is: Whenever alternative actions are possible, select the one which maximizes mathematical expectation.

There are a few problems with this rule. Some claim it over-emphasizes utility and acquisitiveness. Many think that situations exist or can be invented in which the rule violates our rational intuitions. (Should a poor person choose a one-in-a-thousand chance of winning a million dollars over a 90 per cent probability of getting $1,000?) Still, it has been one of the most durable and widely-accepted of all principles of rationality.

The other idea, that a given amount of money is worth less to a rich man than to a poor man, was explicitly stated by Daniel Bernoulli in his attempts to solve the Petersburg Problem. In that instance he formulated a mathematical measure called 'Moral Hope' (or 'Moral Expectation') which purported to show exactly how much a given increase of wealth would benefit an individual with given resources.

Even ignoring the details of D. Bernoulli's argument, it seems intuitively clear to most people that the beggar places greater value on a quarter than the rich man does.[61] We can capture this general notion by saying that the value one places on an increment of wealth is inversely proportional to one's current wealth.

Now let us imagine two persons gambling in a fair game, where each begins with a stake of 100 units. Any outcome except a draw will result in a disparity between the final sums. Suppose A wins 30 units from B. Then A's gain is proportional to 30/130 while B's loss is proportional to 30/70. Thus, *after the game,* the amount transferred looks larger to the loser than to the winner. But it would be irrational to engage in an activity where one risks more than he stands to gain on equal chances. I would not bet my 30/70 against your 30/130 on the toss of a fair coin – that would be unfair odds. But the outcome of any fair game will always exhibit this property, that the loser's relative loss is always greater than his corresponding possible gain. Thus all gambling is shown to be irrational, since most games promise even less than a fair return. As Laplace says[62]

> It results similarly that at the fairest game the loss is always greater than the gain.... We can judge by this of the immorality of games in which the sum hoped for is below this product. They subsist only by false reasonings and by the cupidity which they excite and which, leading the people to sacrifice their necessaries to chimerical hopes whose improbability they are not in condition to appreciate, are the source of an infinity of evils.

This conclusion is much more typical of the Classical theorists than one would have thought. We tend to picture them as eagerly working out the odds in order to excel at the widespread gambling in decadent aristocratic society. In fact, however, they were generally opposed to gambling, treating it only as an item of mathematical interest. Besides Laplace and D. Bernoulli, de Méré and Montmort remarked on the wastefulness of gambling,[63] and Cardano seems to have been the only inveterate gambler of the bunch.[64]

But even if the classical investigation led to the conclusion that gambling is irrational, it was, after all, probability theory which made it possible to demonstrate this. So it has advanced our rationality at least in this respect. (Montmort[65] cites the example of a lottery operator who, lacking a knowledge of probability, offered the public terms unfavorable to himself – alas, we are beyond that now.)

In the end, of course, probability theory has found so many applications that we can scarcely question the rationality of its use. At the time of the CTP, however, the only major practical successes were in gambling and insurance. Otherwise the principal virtue claimed for probability was that it was a codification of common sense, extending what we had always agreed with into areas where things were unclear. Laplace put it thus:[66]

It is seen in this essay that the theory of probabilities is at bottom only common sense reduced to calculus; it makes us appreciate with exactitude that which exact minds feel by a sort of instinct without being able ofttimes to give a reason for it. It leaves no arbitrariness in the choice of opinions and sides to be taken; and by its use can always be determined the most advantageous choice, thereby it supplements most happily the ignorance and the weakness of the human mind...there is no science more worthy of our meditations.

11   CHIEF CRITICISMS

It is well that this section is called 'chief criticisms', since it would require most of the book to attempt to list *all* the criticisms that have been directed at the CTP. We will try to impose some order on this mass by discussing them under the following headings:

1 Criticisms of the Principle of Indifference.
2 The limited application of the CTP.
3 Criticisms of inverse probability.
4 Theoretical ambiguity in the CTP.
5 CTP not tied to the real world of experience.

Criticisms of the Principle of Indifference

This is probably the most popular of our five areas of criticism; virtually every author who has written on the philosophical theory of probability in modern times has taken a shot or two at the Principle of Indifference.[67]

The Principle is variously formulated and variously attributed (principally to Bernoulli and Laplace) but is generally held to be central to the definition of probability and the determination of initial probabilities in the CTP.

Generally speaking, the Classical theorists begin with equiprobable alternatives and define probability as the ratio of those which are favorable to the total set. But, as Reichenbach notes, 'even if the degree of probability can be reduced to equiprobability, the problem is only shifted to this concept. All the difficulties of the so-called a priori determination of probability therefore, center around this issue.'[68]

The principal 'difficulty' with the definition is that it seems to be glaringly circular. After all, 'equiprobable' is normally construed as meaning 'having an equal probability' – but 'probability' is the very word we are trying to define.

Now Laplace can avoid this circularity by trying to give 'equiprobable' some sense of its own, with no dependence on probability.[69] This he does by requiring that the equiprobable alternatives meet two criteria: (1) they must be 'equally possible,' and (2) they must be 'such as we may be equally undecided about in regard to their existence.'[70]

We hasten to add that 'equally possible' likewise must have a sense which is not parasitic on the concept of probability if the original definition is to avoid vicious circularity. It seems that Laplace meant that equipossible alternatives are those which are on the same logical level and can be subdivided in the same ways. Thus Four and Three are equipossible dice throws, while 'Two or less' and 'More than Two' are not, since the latter cases consist of two and four sub-cases respectively.[71] This requirement will be of interest later when we consider alternatives like Six or not-Six, and Red or Blue or Green.

The other requirement, that we be 'equally undecided' about the alternatives, has been a source of even greater difficulty.

To begin with, it is evident that this phrase must not refer to some actual state of indecision, else the CTP would reduce to a psychologistic subjectivism which is clearly not the intent of its framers. Someone might well be undecided about whether a crap-shooter is more likely to roll a Seven or a Three – the point is that he *should not* be undecided. But if the stricture is to refer to some logical or empirical (non-psychological) feature of the situation, what are we to take it to mean?

One possibility is that we might mean that we have *no evidence whatsoever* regarding the possible outcomes. But, as C. I. Lewis suggests, 'it should be doubted whether that kind of case can occur,'[72] since almost everything which can be described has *some* relation to past experience.

A better interpretation is that we are equally undecided about alternatives when they are 'symmetrically related to the body of the evidence.'[73] It would then be all right to know something about the alternatives, as long as that knowledge did not allow us to distinguish between them.

But of course we *can* distinguish between the alternatives: a Five has more pips than a Three, and a Heart and Club have different colors. In fact, Leibniz would tell us that there must be *some* difference between the alternatives, else they would be identical (Principle of the Identity of Indiscernibles).

So our requirement must be amended to allow some differences. This will be all right so long as the differences are not relevant to the problem. But this solution is also unacceptable, because 'if "relevance" is defined in terms of "probable", the circle in the Laplacian definition is once more patent; while if judgments of relevance are based on definite empirical knowledge, the ground is cut from under the basic assumption of the Laplacian point of view.'[74]

The problem is that we want to be able to say that the color of a racing car, for example, is irrelevant to its chance of winning, while the size of its engine is very important. But it should be clear that no general rule can be formulated for distinctions like this, and empiricists would argue that all particular rules are in fact based upon experience rather than a priori logic. The identification of equiprobable cases then seems to be a matter of induction from experience rather than prediction from ignorance.[75] Whenever experience has shown that there are good grounds for treating certain alternatives as equiprobable (as in dice and card games especially) it is perfectly reasonable to use the Principle of Indifference as a rule of thumb for computing initial probabilities. But in such cases the Principle is no longer theoretically fundamental (experience is) and it must be abandoned if future experience gives us grounds for doubting that equiprobability obtains.

In addition to general theoretical questions about the Principle of Indifference, there are also many problems, difficulties, and even contradictions in its application.

Two of the simplest counter-examples turn upon the problem of dividing up alternatives. First, if we are about to throw a die, we might throw a Five or we might not throw a Five. Since we are undecided about the alternatives, the Principle of Indifference might seem to require us to treat them as equally probable. Obviously this criticism turns on the notion of equipossible cases and is invalid *if* the defenders of the Principle succeed in making that concept clear and workable.

The second counter-example is somewhat similar. Suppose we are about to draw a ball of unknown color from an urn with unknown contents. It seems that we have no more reason to expect the ball to be Red than not, so the probability would seem to be 1/2 that it is Red. But of course the same reasoning applies to Green and to Blue, so that $P(\text{Red}) = 1/2$, $P(\text{Green}) = 1/2$, $P(\text{Blue}) = 1/2$. But this gives us a total probability of 11/2 in violation of the basic rule that probabilities must add to 1.

Again the problem seems to be that the alternatives are not equipossible because there are more ways to be non-Red than to be Red. But what if we are asking about the probability that Martians (or other extraterrestrials) are friendly? Are there more ways to be friendly or unfriendly (especially if you're a Martian)? Doesn't it seem that there is equipossibility between these alternatives? And aren't we equally undecided about them? Doesn't it follow from the Principle that they are equiprobable? But then we have the extra-ordinary conclusion that we know the probability that Martians are friendly is exactly 1/2! *Do* we know any such thing?

It seems that equipossibility may eliminate puzzles when we can specify 'ultimate properties' and 'different ways of being X,' but for all those propositions that do not allow such distinctions, if we know nothing about them, then they are just as likely to be true as false! Many have thought this conclusion absurd and a major criticism of the Principle of Indifference.

Besides simple negation, there are other ways of creating alternatives which may or may not be equiprobable. One of the most famous of these is Bertrand's Box.[76]

Consider a Box or chest which has three drawers. We know that one contains two gold coins, one contains two silver coins, and the third contains one gold and one silver coin. Suppose we pick a drawer at random and blindly withdraw a coin which turns out to be gold. What is the probability that the other coin is gold?

Solution I: Since a gold coin was drawn, we must have chosen either the first or the third drawer, but we have no reason to prefer either, therefore the probability is 1/2.

Solution II: We have either drawn the single gold coin in drawer three, or the first gold coin in drawer one, or the second gold coin in drawer one. Of the three possibilities, the latter two are favorable. Therefore the probability is 2/3.[77]

Here we have different applications of the Principle of Indifference, depending on different specifications of the alternatives and resulting in different values for the probability. It is not immediately evident to most people which is the preferred solution (the second) because it is not immediately obvious that the gold coin is a stronger indicator of the first drawer than of the third. This example at the very least shows that much more care must be taken in specifying alternatives than is generally assumed.

An even more complex problem in the application of the Principle of Indifference arises when we consider problems involving continuous variations or 'geometrical' probability.

By far the most famous of these is Bertrand's Paradox. The problem is: for a given circle, what is the probability that a random chord is longer than the side of an inscribed equilateral triangle?

At least three different solutions are possible.

1 If one attends to the end-points of the random chord and computes their possible location, the resulting probability is 1/3.
2 If one attends to the location of the chord's mid-point along the length of the diameter which bisects it, the probability is 1/2.
3 Finally, if one asks whether the mid-point of the chord does or does not fall within a concentric circle of appropriate diameter, the probability seems to be 1/4.

This example is like Bertrand's Box Paradox in that the difficulty turns on the specification of the relevant alternatives. It differs in that there is *no* 'preferred' solution to the problem as stated. However, Kneale (whose discussion of the paradox we have followed here) has pointed out that the problem does become determinate if we specify the method of selecting the 'random' chord. Then we find out that (1) is appropriate for spinning a spinner twice, (2) is appropriate for sliding a ruled glass plate along the surface, and (3) gives the correct result if we let a raindrop determine the midpoint.[78]

So again we have found that a problem can be resolved with more complete information and more careful thought. But now let us consider a paradox which is genuinely irresolvable because it involves attributes which vary continuously. This is the problem of volume and density which Keynes attributes to von Kries.[79]

Suppose we know that the specific volume (volume per unit mass) of a substance lies somewhere between 1 and 3. The Principle of

Indifference then indicates that it is just as likely to be between 1 and 2 as between 2 and 3, so the probability of each of these alternatives is 1/2.

Specific density (mass per unit volume) is the reciprocal of specific volume. Thus our initial condition requires that the specific density lie somewhere between 1 and 1/3. If we apply the Principle of Indifference to equal intervals on this scale, we find that there is an equal probability of 1/2 that the specific density will lie between 1 and 2/3 and that it will be between 2/3 and 1/3. But since the relation is reciprocal, these values correspond to a specific volume ranging between 1 and 1 1/2, and 1 1/2 and 3 with equal probability, contrary to our original calculations.

Other, similar, paradoxes can be generated by the fact that it is not possible for x^2 to be evenly distributed in a domain if x is, and vice versa.[80]

We may say, then, that the Principle of Indifference just will not work reliably on problems involving continuums or an infinity of alternatives – what has traditionally been called geometric probability.

Our final objection to the Principle of Indifference is that we cannot get knowledge from ignorance[81] and no matter how carefully we specify alternatives, 'there is in fact no logical relation between *the number of alternative ways* in which a coin can fall and *the frequency* with which these alternatives in fact occur.'[82]

It seems quite clear that if we are ignorant of the outcome then we are (indeed) ignorant of the outcome and that counting alternatives in such a case will not tell us (in the absence of some information) what the true probability of occurrence of an event will be. But much of the CTP is not directed towards empirical knowledge of events – rather it is intended as a guide to action under conditions of uncertainty. I contend, therefore, that what the Principle of Indifference measures, and measures correctly, is *not* the probability that an event will occur, but rather the probability that a random guess about an event will be correct.

If we arbitrarily bet on a horse in a field of n horses, for example, we should not think that the odds of *that horse winning* are $1/n$, for in general that will be false and certainly its odds of winning are more determined by the quality of horses in the field than by their quantity. What the Principle of Indifference does tell us – and rightly so – is that we should figure the odds that we have picked the winner to be $1/n$, and bet on that expectation. There are two distinct activities here: the horse race, and the random guess. The odds for success in the horse race are determined by such empirical factors as jockeys, track conditions, and equine excellence. But experience has taught us that the odds of success in a truly random *guess* depend only on the number of favorable outcomes compared to the total number of alternatives. Notice that this success ratio is suggested by the phrase 'random guess' with its implication that each alternative stands an equal chance of being chosen. Curiously enough, experience has also taught us that the guess need not be random in this strict sense. When Americans are placed in a situation where they must go to the right or left, they show a marked preference for the right. I do not know (or know if it *is* known) whether this preference is due to the right-handedness of the majority (as seems likely) or the English reading direction (as I have seen suggested) or even the right-wing political orientation of the majority (as my radical friends might conclude, since they think politics determine all human activity). But the important thing is that, in practice, such guesses are successful about 50 per cent of the time even though 'right' is guessed most frequently (just as coin-calls are about 50 per cent successful although calls of 'heads' predominate). It appears that the only requirement we need impose on a random guess is that the principle of selection employed (such as color, identification number, etc.), if any, must have no bearing on the success of the chosen alternative. This definition of 'random' is practically identical to that employed in von Mises's requirement of randomness in a collective. It is generally an *empirical* question whether a guess is random in this sense (perhaps red cars *do* win more races, because people who drive red cars are more reckless than others). Nevertheless, the idea is common enough that most people know what it means to make a random choice in this sense, and, indeed, many of us have developed principles of choice which are *designed* to be random ('Eeenie, meenie, minie, moe...'). I think, therefore, that the concept of *random guess* is an *effective* concept, in the sense that we seldom make mistakes in its application. All our past experience joins with the logical arguments to support the contention that the probability of success in a random guess is given by the Principle of Indifference. I think this is the underlying truth which has accounted for the historical acceptance of that principle. The logical and empirical difficulties which have so discredited the Principle of Indifference are due entirely to the fact that so many have confused the probability of *choosing* a success with the probability of *being* one.

The Limited Application of the CTP

If we are correct in arguing that the Principle of Indifference is primarily adapted to random guessing between clearly defined alternatives, we should not be surprised to find that it can be employed usefully in only a comparatively few situations.

We have already found, for example, that the application of the Principle can be ambiguous and inconsistent in some cases, and that it seems to break down completely in cases involving geometric probability.

Another area where the CTP breaks down, Kyburg points out, is irrational probabilities:[83]

There are also probability problems in physics and mathematics which lead to irrational numbers as probabilities. These probabilities (e.g. $6/\pi^2$ for the probability that an integer selected at random is prime) cannot be regarded as ratios of numbers of alternatives for the simple reason that irrational numbers cannot be regarded as ratios of integers at all.

In addition, our discussion of mortality tables and actuarial problems has shown that interpreting a person as equiprobably represented by each entry in a table is a possible but rather strained way of trying to stretch the theory to fit a desired application. If we have to invent hypothetical charts for all stable frequencies and try to make these charts as complete and comprehensive as possible, our statisticians will be working overtime. And we still have the problem of making sense of mortality charts of future deaths. On the whole, it seems that von Mises was justified in the following criticism of Classical authors:[84]

When the authors have arrived at the stage where something must be said about the probability of death, they have forgotten that all their laws and theorems are based on a definition of probability founded only on equally likely cases. The authors pass, as if it were a matter of no importance, from the consideration of a priori probabilities to the discussion of cases where the probability is not known a priori but has to be found a posteriori by determining the frequency of the different attributes in a sufficiently long series of experiments. With extraordinary intrepidity all the theorems proved for probabilities of the first kind are assumed to be valid for those of the second kind.

Von Mises is perhaps overly snide, since the theorems of the calculus developed by the Classical theorists *do* hold for the frequency interpretation, but he is certainly correct in pointing out the gaping theoretical chasm ignored by the CTP. We might conclude, then, that the solutions worked out in this type of case by the CTP were correct, but unjustified.

It is even clearer that the CTP cannot deal with loaded dice and coins in a way consistent with its theoretical foundations. If we know that a die is loaded in favor of a Five, we must conclude that the alternatives are still equi*possible*, since they are on the same logical level and can be subdivided only in the same way, if at all. But the alternatives are no longer equi*probable*, since we are no longer equally undecided as to which will appear. On a strict interpretation of the CTP (one which takes the Principle of Indifference as the only source of initial probabilities) there is no probability here, since there are no equally probable alternatives. (We shall speak in the next subsection about Laplace's, Bernoulli's, and Bayes's attempts to finesse this difficulty by using inverse probability.)

In fact, any application of probability theory or use of the terms 'probably,' 'probability,' 'chance,' 'odds,' etc., which is not based on equiprobable alternatives is a deviation from the strict CTP. This obviously includes most of our casual talk ('I'll probably fail the exam'), much of our gambling ('The odds on Gluefoot are 20 to 1'), and much of our commercial and scientific activity ('The probable error is 2 per cent'). A strict Classical theorist is left with very little to work on, primarily cards and dice.

Even the application to cases like cards and dice, of course, requires empirical assumptions and hidden rules. In particular, we must assume that the dice are 'fair'. But 'fair' dice are defined as those which give results consistent with the Classical Theory and the Principle of Indifference.[85] It is not surprising therefore – indeed it is analytic – that fair dice act as predicted. The difficult thing is identifying fair dice (or other alternatives) in the real world.

On the whole, there are not many important situations where

equiprobable alternatives present themselves. James Bernoulli recognized this in 1705 when he wrote:[86]

> In the game of dice, for instance, the number of possible cases (or throws) is known.... But what mortal, I ask, could ascertain the number of diseases, counting all possible cases, that afflict the human body in every one of the many parts and at every age, and say how much more likely one disease is to be fatal than another.... These and similar forecasts depend on factors that are completely obscure, and which constantly deceive our senses by the endless complexity of their inter-relationships, so that it would be quite pointless to attempt to proceed along this road.

The other road that Bernoulli opened with his theorem is intended to be a pathway to probabilities that are not based on the identification a priori of equiprobable alternatives, but are instead derived from experience by the method known as inverse probability.

Criticisms of Inverse Probability

Bernoulli's method for establishing probabilities a posteriori is commonly called the Inverse Bernoulli method, or the Inversion of Bernoulli's Theorem, because it is based on that more fundamental principle which bears his name. The basic theorem itself has been the source of a good deal of confusion, so we will try to clear up some of the problems concerning it before we proceed to its inversion.

The central idea of Bernoulli's Theorem is that a repetitive event whose probability of occurrence is p on each of N possible independent occasions will exhibit a frequency of occurrence f that falls in the range $p \pm e$, with a probability of P. The value of P for fixed population N depends on the allowable variation, e; but P is also directly dependent on N, and will continue to approach 1 as N increases. (Bernoulli's Limit Theorem says it can be made to approach closer to 1 than any arbitrary value, d, if the population is increased sufficiently.)

The contested claim for this theorem is that it allows the CTP to predict relative frequencies and to derive probabilities from observed frequencies.

There are two possible situations here: (1) The initial probability p is known, (2) p is unknown.

If p is known, the probability calculus and other general mathematical laws are sufficient to justify the calculation of the final probability. (Some minor restrictions are imposed by the method of approximation actually employed, but these problems are mathematical rather than philosophical.) Unfortunately, not everyone has always remembered that even this final value remains only a probability. There has been a serious misconception from time to time that Bernoulli's Theorem actually established a frequency of occurrence ($f \pm e$) which *will* be observed in experience.[87] We must remember that the theorem is nothing more than a part of the probability calculus, and that it shares with most other parts of that calculus the form 'If probability A is x, then probability B is y.' In the normal application, probability A is the 'initial' probability of a given result in a repeatable empirical trial. Probability B is then the probability that the relative frequency of such a result will fall within a certain range. *But probability B is not itself a relative frequency.* Those who think that Bernoulli's Theorem 'goes from probabilities to frequencies' are therefore mistaken – it goes from probabilities of independent events to probabilities of frequencies of such events.[88]

If this rather elementary error is avoided, there yet remains some question about just how the theorem relates to actual events. Von Mises, for example, criticizes Bernoulli on the grounds that either his theorem is purely arithmetical, and thus about numbers rather than the world, or else it is empirical because of the assumption of the postulates of the frequency interpretation.

He explains the mathematical basis of Bernoulli's Theorem (or the Law of Large Numbers) somewhat as follows. Let us represent the result of a coin toss by '0' for Heads and '1' for Tails. Then the result of a series of 100 tosses corresponds to one member of the set of all 100-digit numbers consisting of only '1's and '0's. There are exactly 2^{100} such numbers.[89]

Considered in this light, Bernoulli's Theorem is seen to have the following arithmetical content:[90]

Let us write down, in order of the magnitudes, all 2^n numbers which can be written by means of 0's and 1's containing up to n figures. The proportion of numbers containing from $0.49n$ to $0.51n$ zeros (assuming $e = \pm 0.1$) increases steadily with an increase in n.

This proposition is purely arithmetical: it says something

about certain numbers and their properties. The statement has nothing to do with the result of a single or repeated sequence of 1000 actual observations and says nothing about the distribution of 1's and 0's in such an experimental sequence. The proposition does not lead to any conclusions concerning empirical sequences of observations as long as we adopt a definition of probability which is concerned only with the relative number of favourable and unfavourable cases, and states nothing about the relation between probability and relative frequency.

Von Mises thus concludes that the arithmetical derivation of this mathematical theorem *has no bearing* on the truth or falsity of Poisson's proposition which he also called the Law of Large Numbers. That is the empirical proposition whose content is roughly that the relative frequencies of certain empirical events tend towards limiting values as the sequence is extended. *This* 'Law of Large Numbers' is not and cannot be proved mathematically. Rather it is the first postulate of von Mises's probability theory.

Finally, von Mises demonstrates that the Bernoulli-Poisson Theorem (the mathematical Law of Large Numbers) indeed *can* be derived in a way that says something about probability and the world, rather than merely number theory, but not so simply as Poisson thought:[91]

The correct derivation of the Poisson Theorem based on the frequency definition of probability requires not only the assumption of the existence of limiting values but also that of complete randomness of the results. This condition is formulated in our second postulate imposed on collectives.

Since this derivation requires the assumption of the postulates of the relative frequency theory of probability, it follows that the Law of Large Numbers cannot be used by Classical theorists as a bridge from their theoretical structure to the use of empirical frequencies in probability calculations.

The essence of this criticism is that if we start out by saying 'There are this many possible ways E can occur...,' we will get from Bernoulli's theorem a conclusion that says 'There are this many possible ways the frequency can occur....' But according to von Mises (and Nagel) such an expression says nothing about how frequently something *will* occur. If we wish this latter (useful) information instead, we must begin with something like '*E* occurs a certain percentage of the time...,' which is not at all a statement of Classical probability. (This point will recur near the end of this section.) Before passing on to the inversion of the Theorem, we will briefly note two fairly common errors which arise in connection with the Theorem itself and which James R. Newman has ably corrected.

The first of these is the belief that the number of successes will get closer and closer to Np as N increases. In fact, the proportionate difference decreases, but the (probable) absolute difference increases. If we go from 1,000 trials to 10,000 trials, it is true that it becomes more probable that the number of successes falls between $0.49N$ and $0.51N$ (say). But the probability that it will fall within $0.50N \pm 10$ *decreases.*[92]

The second, and somewhat similar, illusion is that Bernoulli's Theorem guarantees the doctrine of 'the maturity of chances.' This ancient illusion (sometimes called 'the law of averages') makes gamblers believe that a long run of Heads is more likely to be followed by a Tail than a Head because 'long runs are unlikely' and 'it's got to balance out.' But, as Newman reminds us:[93]

> Bernoulli's Theorem is itself the sole ground for expecting a particular proportion of heads in the cointossing game, and it is an essential condition of the theorem that the trial be independent, i.e., without influence on each other. It is patently foolish, then, to invoke the theorem that sets out from the premise that the probability of a head at every toss is 1/2, to prove that the probability is less than 1/2 after a consecutive run of heads. Yet this is the muddleheaded idea underlying all gambling systems.

Now let us turn to our second possible case, where we are to use the inverse of Bernoulli's Theorem to derive an unknown initial probability. If p is not known, the question arises whether the theorem enables us to calculate either p or P, or both. Clearly P cannot be calculated alone, since it only has a value for a given p and e. However, if we have a given sample, with observed relative frequency G, we can *hypothesize* a value of p, compute e ($e = p - G$) and thus compute a value for P. This tells us what probability there is of observing such a sample if the original probability is p. Classicists, Apriorists, and some Frequentists like Reichenbach take this as an indication of the probability that p is the original probability and accept or reject the hypothesis if P is very high or low. This process is known as the Inversion of Bernoulli's Theorem and is accepted as *meaningful* by all schools.

But all this really tells us is that the observed sequence would have been very improbable, e.g., if the original probability had a certain value. That gives us some reason to *doubt* that the original probability *did* have that value, but it certainly doesn't preclude it.

The usual method is to continue applying the inverse Bernoulli method to different values of p until it is established which value makes P a maximum. (That is, which of the possible values of p makes the observed event more probable than do the others.) This is then said to be the most probable value for p. (For simple cases, the most probable value of $p = G$.)

It is this most probable value which Laplace and Bernoulli sometimes treated as 'the value of the initial probability given by the inverse of Bernoulli's Theorem.' But this is far too strong a construal. Even if this value would make the observed events 'less unlikely,' it may *itself* be sufficiently unlikely to override any presumption in its favor. For example, if we observe 20 tries on a roulette wheel in a reputable casino and find only 6 Reds, Bernoulli's Theorem will suggest that the wheel might be biased, since the hypothesis of fairness would allow only a 0.037 probability of this occurrence, while a bias of 0.30 Red would give 0.192 as the probability of this series, which is greater by a factor of 5. Yet we know it is very unlikely that a reputable casino would stoop to biasing the wheel (especially in this manner, which doesn't stand to benefit them at all), and 'innocent' imperfections or normal wear tend to favor numbers, not colors (which are evenly distributed on the wheel). It is much more reasonable, then, to reject the 'most probable' value selected by Bernoulli's Theorem, in favor of the a priori more probable thesis of a mildly unusual result on a fair wheel.

This kind of criticism is basically the same as that applied to Bayes's Rule by its detractors. It is impossible, they say, to arrive at any mathematically sound value for p unless we know already the a priori probabilities that p takes on this, that, or the other value. Such a case is then mathematically soluble, but it is also extremely rare. In most cases the Bayesians are shooting in the dark by hunch while the anti-Bayesians refuse to do so. Whether this method kills enough bears to justify the waste of ammunition is, I think, the ultimate problem in the justification of Bayesian methods. Like most pragmatic problems, it doesn't admit of a simple solution. Clearly the inverse methods do often give us useful information, and clearly there are situations where it is reasonable to act on their output as the best available guide to action if action is required. Just as clearly, it would be foolish to go around betting that every coin or roulette wheel which shows a slight run must be biased, or that most professors must be Anglicans because two of the three we have met have been.

The methods of inverse probability, like most of the methods of probability theory, are important and valuable because they can sometimes aid our common sense and our intellects in deciding on a course of action. They become pernicious only when they seek to replace, rather than aid, those traditional sources of wisdom.

Theoretical Ambiguity in the CTP

We have already had occasion to mention the difficulty in establishing exactly what the CTP *is*, what its definitions and assertions are, and so on. This difficulty arises partly from the fact that early writers had neither the training nor the inclination to specify exactly what they meant in every case. But, it seems, it is also partly due to the fact that neither the group nor any individual had worked out a systematic and coherent theory of what probability is. Indeed, many of the problems and contradictions which we find in the CTP had not yet been thought of, so it is natural that they should not have been dealt with satisfactorily. We are left, then, with the problem of interpreting their assertions in a way which yields a systematic theory compatible with their uses of 'probability'. It may not be possible to do this because either their assertions or their implied theory may be inconsistent. It also may not be possible to do this because *no* formal theory can account for all the normal uses of 'probability'. (This is a possibility we leave open throughout the book, to be discussed in the final chapter.) It certainly would be unwise and unjust to apply to these pioneers the harsh requirements of logical completeness and consistency we impose on contemporaries. Nevertheless, I think we can identify a definite theory in the Classical writings.

The first problem of interpretation is whether the CTP views probability as subjective and psychological, or whether it finds an objective ground for the concept.

Nagel is inclined to classify the CTP as subjectivist, because of the many references to 'degrees of belief' and 'degrees of certainty.'[94] Carnap, however, gives a much more convincing argument to the effect that the Classical authors never *use* the term 'probability' in the way that we would call subjectivistic, but always assume there is one right answer, valid for everyone (or, at least, for everyone sharing the same information). Nowhere in their writings do they imply or accept the possibility that two equally rational persons, possessed of the same information, might *legitimately* arrive at different probability values. This situation – crucially implied by the subjectivist theory – is completely foreign to the beliefs and methods of the Classical theory. I therefore agree with Carnap that the CTP envisions some *objective* conception of probability.[95]

Having reached this conclusion, we may proceed with Carnap to ask if this objective concept is logical or empirical. (A priori theories are objective and logical, relative frequency theories are objective and empirical. The difference is that the latter views each probability as a contingent feature of the external world, while the former does not.)

The principal arguments for the empirical interpretation are (1) references to 'unknown probabilities,' and (2) the tendency to slide imperceptibly from 'probabilities' to 'frequency of occurrence.'

'Unknown' probabilities are inconsistent with a theory which makes probability a measure of some logical relation between propositions. (We always 'know' these probabilities, in principle, just as we always 'know' whether proposition X implies proposition Y.) Yet Laplace *et al.* frequently speak of unknown probabilities and their most probable values, as if such probabilities were analogous to mass, conductivity, and other unknown empirical properties of things.

I trust that the sections on 'Absolute probability,' 'Physical chance,' 'The metaphysical status of P,' and 'The epistemological status of P' have made clear my position that the classical theorists do not recognize any real, independent, chance in the world. They are determinists. There is no such thing as 'the real but unknown probability that X will occur.' Either X certainly will occur, or it certainly will not. Instead I have argued that there are certain types of descriptions of the world which are often employed and which have strongly associated probability values. If we knew, for example, that one-fourth of the balls in the urn are red, we would know that 'the' probability of drawing a red ball is 0.25. It is my contention that many of the Classical theorists' references to 'unknown probabilities' are just improperly worded references to unknown constitutions and proportions which *are* empirical properties commonly *associated with* certain probability values. But a probability is not a constitution, and therefore talk of unknown probabilities is merely loose and not inconsistent talk.

It is somewhat more difficult to maintain this interpretation in actuarial cases, because these often include no unknown distribution of equiprobable alternatives to be associated with the relevant probabilities. The problem dissolves if we are willing to take seriously Laplace's notion of a perfect table of mortality, since this will represent our unknown distribution. If we remain uneasy about the matter, it is not because of our distrust of unknown probabilities but because it is questionable whether the CTP can properly embrace any frequency-related probabilities at all – known or unknown. *This* difficulty therefore merges into the second one.

There is no blinkering the fact that the CTP just *is* muddled about frequencies. The close empirical association between frequencies and probabilities, together with the enticing prospects of Bayes's and Bernoulli's theorems, led to a confusion of these concepts which is quite natural but also, as von Mises noted, theoretically deplorable.

It may be possible to develop a logical theory of probability which deals successfully and consistently with empirical frequencies. Certainly Keynes and Carnap made great progress in distinguishing the two concepts and clarifying the relation between them. It is tempting to graft their results onto the earlier rootstock of the Classical theory. But just as we should not be too harsh, so also we must avoid indulgence. We therefore absolve the CTP of the charge that it is or (ambiguously) might be subjectivistic, but agree with von Mises and Carnap that it fails to distinguish the logical from the empirical in the crucial area of relative frequencies. It is to that extent reprehensibly ambiguous.

CTP not tied to the Real World of Experience

The final, and, to some extent, all-inclusive, criticism of the CTP is that it is purely a theoretical construction not based on the reality of experience.

It is clear that the greatest achievement of the CTP – the calculus of probabilities – is an abstract mathematical object. Even the controversial theorems of Bernoulli and Bayes are, in their pure form, a part of the mathematical skeleton common to all theories of probability. But the worth of probability theory is not (entirely) based on its abstract mathematical virtues – the role of a theory of probability is to apply those mathematics to the world, and it is here that controversy arises.

We might start by saying that the CTP applies to fair dice, unbiased coins, etc. But, as Ayer pointed out,[96] if we define fair dice as those which give results conforming to the predictions of the CTP, we obviously have an analytic (and therefore circular) proposition. On the standard view, analytic assertions are not about the real world at all, but are matters of logic and definition. We must instead find some non-analytic criteria of application for the theory.

The most famous way of attempting this is to apply the Principle of Indifference. In its unrestricted form, this leads to puzzles, unintuitive results, and even contradictions. Even in its purified form, moreover, the Principle remains subject to the criticism that it either is not tied to experience at all or else it is demonstrably false. Nagel, von Mises, and Reichenbach especially argue that the abstract, logical specification of alternatives does not and cannot determine what will actually happen in experience. (The most obvious example is the biased die, where the alternatives are exactly the same but the frequency of occurrence and therefore (?) the probability are different from the fair die.) Such critics argue that, since indistinguishable alternatives are *not* always equiprobable, we need some other way to tie probabilities to experience.

In our discussion of the Principle of Indifference, we have already covered the problem of the non-empirical origin of Classical probabilities; now we shall examine the charge that they are even further removed from experience because they are unverifiable and unchangeable.

First, it is claimed that Classical probability statements are unverifiable because their definition, and therefore their content, says nothing about the world. To say that the probability of Heads is 1/2 is to make an assertion of one's ignorance of any way to choose between Heads and Tails – it is *not* to say that Heads will come up half the time, so it is irrelevant whether Heads *does* come up half the time or not.

The first rejoinder that springs to mind for most defenders of the CTP is to argue that they do have ways of verifying probability statements – Bernoulli's Theorem can give us strong evidence for or against probability values.

Here I must agree with Nagel that this is not an allowable escape: Bernoulli's Theorem equivocates on 'probability'. The first, or initial. 'probability' is defined in terms of equiprobable alternatives – by the time the final 'probability' is reached it is treated as a frequency (or, perhaps, likelihood) of occurrence. The CTP cannot have it both ways – it must abandon one of these uses or tie them together somehow.

I think the reason the Classical theorists did not see this theoretical error is that, for them, 'probability' and 'likelihood of occurrence' already *were* tied together, semantically if not theoretically. For them it was part of the meaning (perhaps the central part) of 'probability' that it describes something's chance of occurrence. When Laplace put forth his famous 'definition', I contend, he was actually formulating an explicit criterion for probabilities, or a rule for establishing initial probabilities – he was not trying to *define* 'probability' because everyone knew what it meant already, and theorists of that time didn't bother to define words in current usage.

To say all this, however, is merely to excuse the Classical theorists by pointing out that it was a very natural error to make – it does nothing to rectify the error. It is my view that the a priori (logical) school of probability developed in large part as an attempt to purify and correct the Classical theory. Carnap, for example, attacked the verification problem by arguing for logical, rather than empirical, verification.[97] I think that ultimately this is the only way to go for a justification of the Classical position, but it requires extending and modifying the CTP so much that it practically loses its identity and merges into the A Priori Theory. We will therefore stick with the Classical notion that probabilities are verified when events conform to the predictions of Bernoulli's Theorem.

Now for the charge that Classical probabilities are unchangeable. This is of course based on the fact that the definition admits of no way of taking account of future experience. If the die has six sides, the probability of a Five is 1/6 and remains so no matter how often Five shows in the future.

First, a *tu quoque* rejoinder: relative frequency probabilities also 'don't change.' They are fixed and immutable (if they exist). Only our *estimates* of those probabilities can be modified by experience, and the RF theorists imply otherwise only because they have muddied epistemological categories. (More on this in the RF chapters.)

Second, the Classical response turns on the overlooked phrase 'to such as we may be equally undecided about' in Laplace's definition. While this condition is *initially* fulfilled for a biased die, even a brief stretch of experience might make that indecision unequal, and a longer stretch, formalized by Bernoulli or Bayes, might make the evidence overwhelming and even tell us what the true probability is.

Obviously this response is subject to the types of criticism we have raised above. However, it is, I think, what the Classical theorists would say, and it should stand or fall in that respect.

As a final response to this criticism I would like to present a modern example where the Principle of Indifference was first applied, and then revised in its application. The subject of this famous example is the development of the Bose-Einstein statistics. Feller describes it thus:[98]

Consider a mechanical system of r indistinguishable particles. In statistical mechanics it is usual to subdivide the phase space into a large number, n, of small regions or cells so that each particle is assigned one cell.[99] In this way the state of the entire system is described in terms of a random distribution of the r particles in n cells. Offhand it would seem that (at least with an appropriate definition of the n cells) all n^r arrangements should have equal probabilities. If this is true, the physicist speaks of *Maxwell-Boltzmann statistics* (the term 'statistics' is here used in a sense peculiar to physics). Numerous attempts have been made to prove that physical particles behave in accordance with Maxwell-Boltzmann statistics, but modern theory has shown beyond doubt that this statistics *does not apply to any known particles*: in no case are all n^r arrangements approximately equally probable. Two different probability models have been introduced, and each describes satisfactorily the behaviour of one type of particle. The justification of either model depends on its success. Neither claims universality, and it is possible that some day a third model may be introduced for certain kinds of particles.

Remember that we are here concerned only with *indis-*

71

tinguishable particles. We have r particles and n cells. *By Bose-Einstein statistics we mean that only distinguishable arrangements are considered and that each is assigned probability $1/A_{r,n}$*

$$\left[A_{r,n} = \binom{n+r-1}{r} = \frac{(n+r-1)(n+r-2)\ldots n}{1\cdot2\cdot3\ldots(r-1)\cdot r}\ldots \right]$$

It is shown in statistical mechanics that this assumption holds true for photons, nuclei, and atoms containing an even number of elementary particles.

Another model, the Fermi-Dirac statistics,[100] which assumes '(1) *it is impossible for two or more particles to be in the same cell, and* (2) *all distinguishable arrangements satisfying the first condition have equal probabilities*', is found to apply

to electrons, neutrons, and protons. We have here an instructive example of the impossibility of selecting or justifying probability models by a priori arguments. In fact no pure reasoning could tell that photons and protons would not obey the same probability laws.

It is perhaps misleading of Feller to suggest that no theoretical considerations could conceivably have dictated the Bose-Einstein statistics and that they necessarily had to be discovered by a kind of 'cut-and-try' empiricism. Actually there is a way of theoretically accounting for the new statistics. That way is found in a new conception of physical reality which began to develop shortly after Einstein's paper (and partially as a result of it) and came to be known as quantum mechanics. One of the leading proponents of the new mechanics, Max Born, has this to say of Einstein's statistics:[101]

I cannot see how the Bose-Einstein counting of equally probable cases can be justified without the conceptions of quantum mechanics. There a state of equal particles is described not by noting their individual position and momenta, but by a symmetric wave function containing the co-ordinates as arguments; this represents clearly only one state and has to be counted once. A group of equal particles, even if they are perfectly alike, can still be distributed between two boxes in many ways – you may not be able to distinguish between them individually but that does not affect their being individuals. Although arguments of this kind are more metaphysical than physical, the use of a symmetric wave function as representation of a state seems to me preferable. This way of thinking has moreover led to the other case of gas degeneracy, discovered by Fermi and Dirac, where the wave function is skew, and to a host of physical consequences confirmed by experiment.

So, according to Born, the seemingly capricious and inexplicable probability behavior of fundamental particles is in fact a consequence of the theory of quantum mechanics. Of course, even this deducibility does not negate Feller's contention (which I support) that the test of a probability model is its pragmatic success and our inductive conclusion about its future success. But notice that the new statistics, the Bose-Einstein model, is *also* an application of the Principle of Indifference for the computation of probabilities – the empirical question is merely whether state- or structure-descriptions are to be treated as equiprobable.[102] We see then that it is possible to revise or even replace Classical models. They are not as unalterable as critics would have us believe. The critics, however, are right in claiming that the CTP contains no explicit and theoretically satisfactory procedure for effecting such changes.

12   CHIEF VIRTUES

I would like to speak first of the virtues of the founders of the Classical Theory of Probability. We must remember that they were not just developing *a* theory of probability, they were making the very first efforts to treat probability in a rigorous mathematical fashion. It is true they made a few blunders and left us a few theoretical muddles, and one can even argue that their basic rule, the Principle of Indifference, is unacceptable. But these quibbles and shortcomings are far outweighed by their astonishing accomplishments. Remember that the calculus of probability, the fundamental mathematical structure of all probability theory, was created *ex nihilo* by these thinkers. Their brilliance is attested by the fact that this calculus remains virtually unchanged to the present day, and is still accepted by every major theory of probability.

But if the Classical theorists' chief claims to fame are as trailblazers and as the developers of mathematical methods, we must hasten to add that their theory hasn't done all that badly either. Its two great virtues are its simplicity and its pragmatic success. Undoubtedly more people have used its methods than those of all the other theories combined. (Put another way, there are more card players (CTP) than there are logicians (AP), scientists and actuaries (RF), and psychologists (SUB) combined.)

Whenever we are interested in cases that have readily distinguishable alternatives that *are* equiprobable, the CTP readily gives answers of immediate practical value. Furthermore, if one is not bothered by theoretical niceties, it is possible to follow the Bernoullis, Bayes, and Laplace into ever-widening areas of application that also allow pragmatic success, even if they depart somewhat from the pure definition.

Indeed, as the Bose-Einstein example shows, a creative reinterpretation of the basic alternatives can often change failures into successes. Finally, we should note the extraordinary success and longevity of the CTP's definition of probability. Subsequent theorists have attacked the CTP vigorously and offered various alternative theories. Kyburg has even claimed that the Classical theory has 'long been abandoned among philosophers and reflective mathematicians.'[103] This may well be so. But evidently not everyone falls in these exalted categories, for consider these recent definitions:

> *Probability.* The ratio of the number of ways in which an event can occur in a specified form to the total number of ways in which the event can occur.[104]
> *Probability, Mathematical.* If an event can happen in a ways and fail in b ways, and, except for the numerical difference between a and b, is as likely to happen as to fail, the mathematical probability of its happening is $a/(a + b)$ and of its failing, $b/(a + b)$.[105]

The Classical definition lives! And as long as human beings continue to face situations where the outcome is unknown but the alternatives are all felt to have an equal chance of occurrence – as long, in short, as we continue to gamble at dice and cards – the Classical Theory of Probability will continue to be the working theory of the ordinary person.

III

A PRIORI THEORIES
OF PROBABILITY

In this chapter we will discuss a priori (AP) theories in general, with exemplary excursions into particular theories. We will be concerned with (1) basic ideas of AP theories, (2) chief criticisms of such theories, and, finally, to end on a positive note, (3) chief virtues of AP theories. We will deal primarily with the theories of Keynes and Carnap, but, in the main, our remarks are applicable to most other AP or 'logical' theories (Jeffreys, Koopman, Kemeny, Hintikka, etc.) as well.

1   BASIC IDEAS OF A PRIORI THEORIES

It seems to me that the basic ideas which all AP theories of probability share are chiefly three:

1 Probabilities are known (or determined) a priori, not by (purely) empirical means.
2 Probability is a logical relation between sentences (propositions, events, properties, predicates).
3 A probability is always relative to given evidence only.

The first point is of course the most important characteristic of a priori theories and the one which gives them their name. It is what distinguishes them most strongly from the relative frequency or 'empirical' school.[1] It is also what occasions the single greatest objection to them: How can a priori principles do probability theory's practical work of predicting the future in the real world? Again, it is the source of their greatest single advantage – they can establish probabilities without the necessity of waiting for a very (infinitely?) long sequence of repetitive empirical events.

The second of our basic ideas, like the first, is so central to AP theories that it has generated a name for them: 'Logical Theories of Probability.' As interest has grown in the inductive logics of Carnap, Kemeny, Hintikka, *et al.*, this term has been more and more commonly used for such theories. I agree that it is a suitable, descriptive name for this sub-class of a priori theories. I object, however, to its extension to theories of the Keynesian variety, because there the initial probabilities *are not* obtained from quantitative logistic systems but from a priori intuition aided by the Principle of Indifference.[2] My objection is not fervid, however, and if the philosophical tide continues to flow towards empiricism in general but away from the dogmas of logical positivism, I foresee a time when 'a priori' will become a general pejorative and I too shall abandon it. Of course the important point here is that these theories *do* define probability as a logical relation – not whether that fact is sufficient to warrant a label.

This concern with logic was motivated in Carnap's case, apparently, by a conscious desire for an inductive logic (more than by a need to account for the probability calculus, for example). He had earlier accepted RF theories as adequate for scientific probability – not until he engaged in logical research did he shift his allegiance to AP theories (or, rather, divide it between the two). Keynes, on the other hand, seems to have set out looking for the nature of probability and then 'discovered' that it is a matter of logic.

The two also disagreed on the relation between deductive logic and probability. Carnap took the (normal) position that probability theory results from adding probability rules (*c*-functions) to the ordinary deductive logic, thereby increasing its power and range. Keynes, however, thought probability theory was the basic theory of inference – deductive logic is merely that degenerate case where all probabilities are 1 or 0.

Our third basic idea is that AP theories recognize probabilities relative to given evidence only. Many RF theorists consider this to be a grave defect in AP theories, rendering a probability subordinate to the state of our knowledge and therefore odiously 'subjective'. Their theories, they maintain, deal with the *real* probability, which is objectively determined and not relative to anything. There is a sense in which this claim for RF theories is true and a sense in which it is false – we shall discuss this problem in the chapter on Relative Frequency theories below. Our present concern is with the AP theories, and in this context evidence-dependence is seen as a virtue rather than a defect. It is a necessary consequence of the fact that probability is defined as a relation between a proposition and some evidence for that proposition.

2 CHIEF PROPONENTS

There are many people who have proposed or are still proposing theories of this general type. I have chosen to concentrate on Keynes and Carnap because they are the intellectual giants of the group and their theories are the most seminal, philosophical, wide-ranging, and fully developed.

John Maynard Keynes (1883–1946), British economist and man of letters, was 'one of the creators of the modern world.'[3] His theoretical contributions to political economy are familiar to everyone as an important part of the *rationale* for the ever-increasing government intervention in non-communist economies. His practical contributions to the British Treasury and to international monetary conferences such as Bretton Woods are well known to at least economists and historians. What is less well known is that philosophy was Keynes's earliest love. He studied it under G. E. Moore and Bertrand Russell as an undergraduate at Cambridge while earning his degree in mathematics. During his subsequent employment with the India Company, Keynes combined these fields in a thorough review of probability and induction. This resulted in a dissertation which earned Keynes a fellowship in King's College and allowed him to return to Cambridge as a philosopher. He was persuaded to teach in the Economics faculty instead, and subsequently made that field his chief intellectual interest, so his brief career as a philosopher culminated in the publication of an expanded version of his dissertation as *A Treatise on Probability*.[4] This one work established Keynes as *the* authority of his time on probability theory, and maintains his reputation today as a leading spokesman for the a priori (AP) interpretation of probability.

Keynes was the first self-conscious apriorist in probability theory. He asserted emphatically that probability is a logical relation between propositions and, according to Carnap, was the first to

77

perceive and emphasize the fact that such probabilities are inherently relative to given evidence and to nothing else.[5]

Such assertions, and the philosophical argumentation which supports them, constitute Keynes's most famous and significant contributions to the development of probability theory. His derivation of theorems in the probability calculus represented at best a mild improvement over his predecessors and is not historically important. His attempted revision of the Principle of Indifference is frequently referred to and generally considered to be an improvement, but it does not attack the fundamental question whether *any such* principle can be a legitimate source of initial probabilities. Thus Keynes is seldom cited for these achievements. Instead, he finds his place as *the* pre-eminent representative of a priori probability theory, and when he is discussed by philosophers it is usually in this role as the embodiment and spokesman of a priori probability rather than as the source of some particular argument or theorem (in contrast to Bernoulli or Bayes, for example).

Rudolf Carnap (1891–1970) was 'the most prominent representative of the logical empiricist, or logical positivist, school in the philosophy of science and logic.'[6] His early training in physics led him to accept the relative frequency (RF) theory of probability,[7] but his name is now more associated with his later development of AP probability theory as a form of quantitative inductive logic.

Carnap's AP theory is of a piece with his notable work in the related fields of logic, syntax, semantics, and formal languages. It is impossible to convey fully the nature of his system of a priori probability without requiring or providing some grounding in the concepts and formalizations he employs. At least a hundred pages of his *Logical Foundations of Probability*[8] are devoted to just this preliminary spadework – the task is obviously beyond us here. Therefore we shall barely sketch in the particular system he develops and concentrate primarily upon his philosophical discussions of the nature of probability theory in general and the justification of his own theory in particular.

3   DEFINITION OF PROBABILITY

In developing a definition of probability, Keynes began with the conscious fundamental principle that probability is a logical relation between propositions and systematically discussed the justification for and consequences of this view. This imbued his work with a unity and philosophical cohesiveness not found in his predecessors. We shall begin by quoting his own statement of his basic position:[9]

> The terms *certain* and *probable* describe the various degrees of rational belief about a proposition which different amounts of knowledge authorise us to entertain. All propositions are true or false, but the knowledge we have of them depends on our circumstances; and while it is often convenient to speak of propositions as certain or probable, this expresses strictly a relationship in which they stand to a *corpus* of knowledge, actual or hypothetical, and not a characteristic of the propositions in themselves. A proposition is capable at the same time of varying degrees of this relationship, depending upon the knowledge to which it is related, so that it is without significance to call a proposition probable unless we specify the knowledge to which we are relating it.

Here we have at once the two themes most frequently associated with Keynes's name: that probability is a logical relation between propositions, and that all probabilities are relative to given knowledge. These must not be confused with the *subjectivist* thesis which had frequently appeared in earlier works on probability (and which led to Nagel's construal of the essence of Classical probability) to the effect that probability is a measure of *our* partial belief in a proposition and that all probabilities are relative to *our* knowledge. Keynes is opposed to such psychologisms (though he occasionally slips into one himself) and sees the probability-relation as an *objective* logical relation. 'A proposition,' he says, 'is not probable because we think it so.'[10] Rather it is probable or improbable with respect to given evidence *whether we recognize it or not*, depending only on whether the logical relation obtains or not.

Keynes symbolizes this logical probability-relation between a proposition a, and evidence (or hypothesis) h, as a/h. He prefers this symbol over the more traditional 'P', because

> The value of the symbol a/h, which represents what is called by other writers 'the probability of a,' lies in the fact that it contains explicit reference to the *data* to which the probability relates the conclusion, and avoids the numerous errors which have arisen out of the omission for this reference.[11]

The probability symbol is *allowed* to take on numerical values and enter into mathematical formulae, so that it is *meaningful* to say, for instance, '$a/h = x$,' or '$3(a/h) = 3x$.' But Keynes does not *require* that every probability possess a definite numerical value. He argues that not all probabilities are *measurable* and not all pairs of probabilities are *comparable*. In his system, it is not true that the numerical values of some probabilities are merely unknown. Instead, he contends that when we speak of unknown probabilities we really refer to the values which we would arrive at *if* we had more evidence or *if* we were more skilled at computation; but by his rules those values are *different* from the present probability. This is so because Keynes thinks the present probability is the result of applying the principles of *human* rationality to the *present* evidence. If either of these factors were to be changed, he contends, we would be talking about a different probability-relation. Also, since the rules and the available evidence do not generate numerically definite probabilities in many cases, we must conclude that some probabilities are non-measurable and non-comparable.[12] The expression 'a/h' can therefore not generally be treated as the name of a number subject to such mathematical laws as are based on the comparability of numbers.

Thus far, Keynes has told us that probability is a logical relation and introduced a symbol for it. Now if we knew *which* relation it is and how to decide its value,[13] we would have a working definition of 'probability.' But Keynes not only fails to help us in this, he denies that it can be done:[14]

A *definition* of probability is not possible, unless it contents us to define degrees of the probability-relation by reference to degrees of rational belief. We cannot analyse the probability-relation in terms of simpler ideas. As soon as we have passed from the logic of implication and the categories of knowledge, ignorance, and rational belief, we are paying attention to a new logical relation in which, although it is logical, we were not previously interested, and which cannot be explained or defined in terms of our previous notions

Even though Keynes tosses aside 'degree of rational belief' as an apparently unworthy definition, in the remainder of his work he treats 'degree of *justified* rational belief' as being, if not equivalent, at least co-extensive (cause and effect, perhaps, or ground and consequent). This must suffice, in lieu of a definition, as an indication of the qualitative nature of the probability-relation: it is a logical or epistemological (certainly non-empirical) quality possessed by two propositions (or sets of propositions) whereby the second warrants some degree of rational belief in the first. Now let us consider what *are* these degrees of rational belief, and how are they determined? By learning to *apply* the probability-relation we can hope to gain further insight into its nature.

Keynes has said already that 'The terms *certain* and *probable* describe the various degrees of rational belief about a proposition which different amounts of knowledge authorise us to entertain.'[15] As we might have expected, certainty represents one extreme on the scale; the other extreme is called 'impossibility.'[16] To maintain Keynes's consistency, we obviously must interpret 'impossibility' in the logical sense of 'self-contradiction' or the epistemological sense of 'negative certainty' – it is not physical, scientific, or empirical impossibility which is involved here. (This point will occur later, as part of a basic criticism of AP theories.)

The extremes are familiar enough. The middle ground is the peculiar province of probability – as opposed to deduction – in Keynes's scheme of logical relations. How, then, do we assign values to the middle ground?

According to Keynes, our knowledge of the middle ground is not epistemologically different from our knowledge of the extremes – in both cases we have 'direct knowledge' which is based on 'direct acquaintance' with logical relations.[17]

Keynes says of probability relations what so many have said of deductive logical relations, that they are a fundamental source of our knowledge, directly available to our intuition, which neither can nor should be referred to anything else as their source or justification. We shall return to this epistemological theory later in this chapter,[18] for now we are concerned with the question of the source of initial probabilities in Keynes's theory, and we will accept 'intuition' or 'direct knowledge' as the *general* answer to that question.

But if we ask the more specific question, 'What is the source of the numerical values which are operated on by the probability calculus?' we receive a more detailed and intriguing answer: the Principle of Indifference.

It is here that Keynes most clearly illustrates the true continuity of probability theory running from the Classical theories through his own:[19]

> It has always been agreed that a numerical measure can actually be obtained in those cases only in which a reduction to a set of exclusive and exhaustive *equiprobable* alternatives is practicable.

Thus Keynes *accepts* the classical First Principle that numerical probabilities are *only* obtainable as the ratio of favorable cases to all equiprobable cases. He feels that his predecessors had not erred in taking the Principle of Indifference as the fundamental source of initial probabilities; rather their errors had been that (1) they had failed to formulate the Principle correctly, complete with the restrictive conditions of relevance and symmetry, and (2) they had sought to apply the Principle in many cases where it was inapplicable.

The fact that the Principle of Indifference is not universally applicable Keynes takes as a consequence and reinforcement of his independent arguments that not all probabilities are numerical (they *would* be numerical, if the Principle of Indifference could be used in every case). He thinks that he has rehabilitated the Principle by requiring that the alternatives be ultimate and symmetrically related to all relevant evidence. Rather than stop to criticize his formulation I will only note that I share Lewis's suspicion (which we shall discuss later) that if one knows enough to be sure that these conditions are fulfilled, one most likely has sufficient empirical evidence to assign the probability on some basis other than indifference. In his review of Keynes, Lewis said only this,[20]

> The reappearance of that *bête noir* of clear thinking, the Principle of Indifference, comes as something of a surprise. But the treatment given obviates the worst objections. Whether it obviates them all is a complex and difficult question.

The basic charges against the Principle of Indifference are that logic alone (without evidence) can say nothing about our contingent world (while some probability statements clearly do) and that no justification is given or possible for the assumption that the members of a logical division are equally *likely to occur*. Such serious challenges are raised by Reichenbach, Nagel, and Lewis and will be discussed later. They are mentioned here only to show that Keynes has *at best* achieved a technical revision which eliminates (at least some) logical contradictions and other absurdities from the implications of the Principle – he has *not* fundamentally altered its theoretical position as the source of all numerical probabilities nor justified its basic presuppositions.

Carnap's definition of a priori probability follows closely the basic Keynesian insight that probability is a logical relation between assertions. In the spirit of modern logic, however, he refined this definition to make probability a mathematical relation between sentences in a formal language. And while he did most of his important foundational work on this theory, he did not think it exhausted the field of probability theory.

We have already mentioned the fact that Carnap explicitly recognizes the difference between the types of theories of probability. A priori (AP) theories he calls 'probability$_1$'; relative frequency (RF) theories 'probability$_2$.' What makes him unique among the major theorists is that he not only recognizes both, he embraces both. He believes that the two theories of probability address themselves to different explicanda; thus he is able to accept each as valid without striking a direct contradiction. This would seem to justify one in classifying Carnap either as an apriorist or a frequentist or both. (In fact, Nagel does include Carnap in a list of frequentists.[21] This was correct at the time, 1939, but Carnap's major work came later and has made reclassification necessary.) Nevertheless, I feel it appropriate to include him in this discussion of AP theories while ignoring him in the RF sections, because Carnap himself emphasized the former aspect of probability over the latter, and because his AP theory has been far more fecund in leading to subsequent developments in probability and inductive logic.

Carnap's principal work in the field – *Logical Foundations of Probability* – first appeared in 1950. His original intention was to produce a two-volume work: the first to deal with probability and establish a suitable system or theory; the second to use that theory as a foundation for the development of a system of inductive logic. The first volume concentrated on AP probabilities because Carnap thought this to be the only acceptable foundation for inductive logic. This preference, which runs throughout the book, appears in the very first paragraph of the Preface:[22]

The theory here developed is characterized by the following

basic conceptions: (1) all inductive reasoning, in the wide sense of nondemonstrative reasoning, is reasoning in terms of probability; (2) hence inductive logic, the theory of the principles of inductive reasoning, is the same as probability logic; (3) the concept of probability on which inductive logic is to be based is a logical relation between two statements or propositions; it is the degree of confirmation of a hypothesis (or conclusion) on the basis of some given evidence (or premises); (4) the so-called frequency concept of probability, as used in statistical investigations, is an important scientific concept in its own right, but it is not suitable as the basic concept of inductive logic; (5) all principles and theorems of inductive logic are analytic; (6) hence the validity of inductive reasoning is not dependent upon any synthetic presuppositions like the much debated principle of the uniformity of the world.

Following Carnap's lead, we shall also concentrate on his views concerning AP probability. These views are first introduced as being part of an historical school of thought:[23]

My conception of logical probability (called 'probability$_1$' in this book) has some basic features in common with those of other authors, e.g., John Maynard Keynes, Frank P. Ramsey, Harold Jeffreys, Bruno De Finetti, B. O. Koopman, Georg Henrik von Wright, I. J. Good and Leonard J. Savage, to mention only the names more widely known. All these conceptions share the following features. They are different from the frequency conception ('probability$_2$' in this book). They emphasize the relativity of probability with respect to the evidence. (For this reason, some of the authors call their conception 'subjective'; however, this term does not seem quite appropriate for logical probability....) Further, the numerical probability of an unknown possible event can be regarded as a fair betting quotient. And, finally, if logical relations (e.g., logical implication or incompatibility) hold between given propositions, then their probabilities must, according to these conceptions, satisfy certain conditions (usually laid down by axioms) in order to assure the rationality of the beliefs and the actions, e.g., bets, based upon these probabilities.

There follows a considerable discussion of how various views are

properly categorized as AP or RF, objective or subjective. Then
Carnap begins to explain[24] his probability$_1$ on the basis that

> The probability$_1$ of a hypothesis *h* with respect to given evidence *e* represents
> (A) a measure of the evidential support given to *h* by *e*;
> (B) a fair betting quotient;
> (C) an estimate of relative frequency.

(A) is a concept which is primarily due to Keynes. (B) had been
central to probability theory since the classical school became
interested in games of chance. (C) represents a new way of looking
at the matter, as well as a potential connection between AP and RF
probabilities similar to the estimate theory of probability which
Lewis had 'developed independently'[25] four years earlier. But there
is some reason for holding that (C) is not fundamental to Carnap's
view at all, because in many cases where Carnap speaks of probability$_1$ being the estimate of a relative frequency, he uses 'estimate'
in a technical sense which he explicitly defines:[26]

> The *estimate* (more explicitly, the probability$_1$-mean estimate) of
> the unknown value of a magnitude with respect to a given
> $e = {}_{df}$ the probability$_1$-mean, that is, the sum of the products
> formed by multiplying each of the possible values of the
> magnitude with the probability$_1$ of its occurrence with respect
> to *e*.

Now this fixed meaning of 'estimate' gives us a definite rule for
arriving at a probability in the sense of 'estimate of a frequency,'
which would seem to make it simple to calculate probabilities. Indeed,
the mathematical manipulation required is precisely the same as that
employed in computing the familiar 'mathematical expectation' of
an event – what could be simpler? The catch, of course, is that
calculation of the estimated value of the frequency requires *prior*
knowledge of the probability$_1$ for each possible value of the
magnitude. But once we have *this* information prediction of the
probable value of the magnitude is a mere exercise in probability
calculus. What is crucial is how the *initial* probabilities are obtained,
and in this procedure it is *not* done by *estimating* a frequency, but
by a logical computation within a formal language. We therefore
conclude that the estimate rule is not fundamental[27] to Carnap's
theory and need not detain us here. (We shall, however, later discuss

some other of Carnap's methods for predicting frequencies which do not depend upon prior probabilities – these are the methods of statistical inference.)

We noted above that an adequate description of Carnap's system is too large a task to attempt here. What we will try to accomplish, however, is to display enough of the nature of his system of probability$_1$ so that its fundamental philosophical assumptions are clear.

According to Carnap, probability$_1$ is a measure of the partial inclusion of the range of one sentence in that of another. Thus it resembles probability$_2$ in being 'the ratio of partial inclusion of one class in another.' That is why the concepts are so similar and so often confused. Yet they are far from the same:

> there remains this fundamental difference: for probability$_2$ the partial inclusion is a factual matter, and hence the value of probability$_2$ is established empirically; on the other hand, probability$_1$ concerns partial inclusion of ranges, which is of a purely logical nature.[28]

Let us set aside some of Carnap's technical machinery by making the rough and ready translation 'The range of a sentence is the number of possible worlds in which it is true' (after a language is specified, of course). The idea is that, given adequate descriptions of the possible worlds, only a semantic, non-empirical investigation is required to establish the range of a sentence. Now if we follow Carnap's proposal that the confirmation (probability) function be the ratio of some measure of the worlds in which both evidence and hypothesis hold, divided by the measure of those in which the evidence holds, we get this result: *Probability is the ratio of the measure of favorable cases to the measure of all possible cases.* This last, of course, is almost identical to the traditional concept of probability, as developed by Laplace and Bernoulli and transmitted by Keynes. Carnap's theory differs from the Classical theory, though, in that measure-functions, and hence confirmation-functions, assign a 'weight' to each possible world, and such weights are not necessarily equal. In the end, however, the particular measure-function, m^*, and confirmation-function, c^*, which Carnap prefers are based on a form of equiprobability.

In *Logical Foundations of Probability* (referred to hereafter as *LFP*) Carnap developed his system for a restricted class of very simple languages, which he calls the languages L. These consist of a finite number of independent one-place predicates (naming properties) applied to a specific number of individual constants (naming individuals) or variables, and the usual logical connectives. If we use 'π' to symbolize the number of predicates (an unfortunate but now well-established convention of Carnap's) and 'N' for the number of individuals, the individual languages receive names of the general form 'L_N^π'. Since predicates and individuals are independent and logically indistinguishable, it doesn't matter for our purposes *which* 2 predicates and 4 individuals occur in a language L^2_4 – the logical, inductive, and probabilistic features of any such language are identical. In fact, all *finite* languages behave similarly – it is only when we allow an infinite number of individuals (in the languages L_∞^π) that we have to beware of logical peculiarities.

Now consider the simple language L^2_2. We will use the two predicates 'M' and 'N,' and the two individual constants 'a' and 'b,' and develop Carnap's system for this elementary case.

The fundamental concept of Carnap's inductive logic is the notion of a *state-description*. A state-description is a sentence (or class of sentences, for L_∞^π) which completely describes a state of the world by affirming or denying each property of each individual. In Carnap's words:[29]

A state-description for a system L in the sense indicated must state for every individual of L and for every property designated by a primitive predicate of L whether or not this individual has this property.

It is heuristically helpful to think of these state-descriptions as characterizing *possible worlds*. A world in which a is M would be different from a world in which a is $- M$. Similarly for all possible combinations of individuals and properties. For any finite system it is always possible (in principle) to list many state-descriptions. Here is such a list for our L^2_2.

1 *Ma & Na & Mb & Nb*	9 *– Ma & Na & Mb & Nb*
2 *Ma & Na & Mb & – Nb*	10 *– Ma & Na & Mb & – Nb*
3 *Ma & Na & – Mb & Nb*	11 *– Ma & Na & – Mb & Nb*
4 *Ma & Na & – Mb & – Nb*	12 *– Ma & Na & – Mb & – Nb*
5 *Ma & – Na & Mb & Nb*	13 *– Ma & – Na & Mb & Nb*

6 Ma & $-Na$ & Mb & $-Nb$	14 $-Ma$ & $-Na$ & Mb & $-Nb$
7 Ma & $-Na$ & $-Mb$ & Nb	15 $-Ma$ & $-Na$ & $-Mb$ & Nb
8 Ma & $-Na$ & $-Mb$ & $-Nb$	16 $-Ma$ & $-Na$ & $-Mb$ & $-Nb$.

We shall use 'Z' as the general name for state-descriptions, and 'Z_i' as the individual names (so that 'Z_1' names the first item in our list, 'Z_2' the second, etc.).[30]

We next introduce the notion of the *range* of a sentence. If h is a sentence of L, the range, R, of h is the class of all state-descriptions, Z, in which h holds. If h is Ma & Na in our example, then $R(h) = \{Z1$ & $Z2$ & $Z3$ & $Z4\}$. The meaning of a sentence is determined by its range,[31] and deductive logic is construed as part of the rules of ranges. ((Ma & Na) $\supset Na$, for example, means that $R(Ma$ & $Na)$ is a subclass of $R(Na)$, which is $\{Z1$ & $Z2$ & $Z3$ & $Z9$ & $Z10$ & $Z11$ & $Z12\}$.)[32] Using these two concepts, the simplest course would seem to be to define the probability of h on evidence e in terms of a ratio between the number of Zs in the range of h & e and the number of Zs in the range of h alone. This is the basic idea that was obscurely suggested by Wittgenstein in the *Tractatus*[33] and serves as the common notion in what Von Wright[34] calls the 'spielraum' (= 'range') school of probability theorists.

Unfortunately, this approach leads to the undesirable result that inductive logic would not allow us to learn from experience. To see that this is so, we will develop this method first, and then the method Carnap is inclined to adopt.

To begin with the general case, the degree of confirmation, q, of a hypothesis, h, on evidence, e, will be symbolized as '$c(h, e) = q$.' Since our fundamental idea is to compare the weight of h & e to the weight of e (favorable cases to all possible cases, given that e is true) we define[35]

$$c(h, e) = \frac{m(e \ \& \ h)}{m(e)}$$

where m is an unspecified function giving the weight of a sentence. Thus specification of a confirmation function (called a 'c-function') is achieved by the prior choice of a means of weighting sentences, which Carnap calls a 'measure-function' (or 'm-function,' for short).

Returning now to the particular case, the first and obvious suggestion is that the weight of a sentence should be the proportion of possible worlds in which it is true. In Carnap's system, this means that the measure of h would be the number of state-descriptions in which it is true, $z(h)$, divided by the total number of state-descriptions, z, in L:[36]

$$m^\dagger(h) = \frac{z(h)}{z}.$$

The confirmation-function based on this measure-function would be:

$$c^\dagger(h, e) = \frac{m^\dagger(e \ \& \ h)}{m^\dagger(e)}.$$

If h is Na and e is $Ma \ \& \ Na$, as above, $m^\dagger(h)$ is equal to the number of Zs in the range of Na divided by z

$$m^\dagger(h) = \frac{z(h)}{z} = \frac{8}{16} = 1/2.$$

Similarly,

$$m^\dagger(e) = \frac{z(e)}{z} = \frac{4}{16} = 1/4$$

and

$$m^\dagger(e \& h) = \frac{z(e \& h)}{z} = \frac{4}{16} = 1/4.$$

Finally,

$$c^\dagger(h, e) = \frac{m^\dagger(e \& h)}{m^\dagger(e)} = \frac{1/4}{1/4} = 1.$$

This result is very satisfactory: since $Ma \ \& \ Na$ implies Na, we naturally wish $c(Na, Ma \ \& \ Na)$ to be 1, but of course, giving satisfactory results for deductive relations is hardly sufficient for a function to be used in inductive logic. So now let's try some induction.

Suppose we are trying to predict whether the individual b will be both $M \ \& \ N$. If we have no empirical evidence at all, the best we can do is to determine the null (a priori, antecedent, initial) confirmation based on logical ranges alone. Let $Mb \ \& \ Nb$ be h. The evidence, *e*, in this case is replaced by the object-language symbol '*t*' (for 'tautology') which holds in every *Z*. Thus we have

$$c^\dagger(h, e) = c^\dagger(h, t) = \frac{m^\dagger(t \& h)}{m^\dagger(t)} = \frac{m^\dagger(h)}{1} = \frac{z(h)}{z} = \frac{4}{16} = \frac{1}{4}.$$

Now suppose we acquire some evidence, *e'* which tells us that the other object in our universe, *a*, is both *M* and *N*. In this case

$$c^\dagger(h, e') = \frac{m^\dagger(e' \& h)}{m^\dagger(e')} = \frac{z(e' \& h)/z}{z(e')/z} = \frac{z(e' \& h)}{z(e')} = \frac{1}{2}.$$

But this is the same value we had before, so our evidence has not changed our assessment of the odds in the slightest.

This result holds for all systems, not just our simple $L^2{}_2$. Even if we had found 100 individuals possessing *M* & *N* with no exceptions, the probability that the 101st would be *M* & *N* would remain equal to the original value of 1/4. In general, then, $c^\dagger$ is defective as a confirmation-function because it fails to learn from experience in a very important way.[37] Carnap has rejected $c^\dagger$ for this reason and constructed another confirmation-function, c^*, which we will consider now.

The new *c*-function, c^*, no longer considers all state-descriptions to be created equal. Instead, it introduces a definite bias towards uniformity by favoring more homogeneous state-descriptions with greater weight than that given to hodgepodges, potpourris, and random collections. To accomplish this, Carnap first introduces the notion of a *structure-description*. The technical definition is:[38]

> *j* is *the structure-description corresponding to* Z_i (or, Z_i *belongs to the structure-description j*) in $L_N =_{df} Z_i$ is a *Z* in L_N, and *j* is the disjunction of all *Z* which are isomorphic to Z_i arranged in lexicographical order.

Two *Z*s are isomorphic if and only if one can be derived from the other by merely exchanging some individuals for others by means of a one-to-one correlation. Our *Z*3 and *Z*9 are isomorphic, because if '*a*' and '*b*' are interchanged in *Z*3, and the result reordered according to the (assumed) lexicographical rules of the system, the result is identically *Z*9. These state-descriptions exhibit a similar *structure*, because each consists of one individual which is both *M* & *N*, and another which is *N* but not *M*. Thus we can specify a structure-description according to the method of the definition:

$$STRx = Z3 \vee Z9$$

or by naming any one state-description in the STR

the STR corresponding to $Z3$

or by specifying *how many* individuals exhibit each possible combination of properties

one M & N and one $- M$ & N.

We can think of these as alternative ways of specifying the structure (arrangement, pattern) of the universe, since they specify how many of each type will exist, but not who will be what. (As 'The club will have one President, two Vice Presidents, and a Secretary-Treasurer' specifies the structure of an organization.)

Now the idea of c^* is to treat each of these structures as equiprobable. In our $L^2{}_2$, the number, z, of state-descriptions is 16, but the number, T, of structure-descriptions is only 10.

$STR1 = Z1$	$STR6 = Z7 \vee Z10$
$STR2 = Z2 \vee Z5$	$STR7 = Z8 \vee Z14$
$STR3 = Z3 \vee Z9$	$STR8 = Z11$
$STR4 = Z4 \vee Z13$	$STR9 = Z12 \vee Z15$
$STR5 = Z6$	$STR10 = Z16$

If we treat each of these structure-descriptions as equiprobable, then the measure of each $= 1/T = 1/10$. If m^* (STR1) $= 1/10$, then $m^*(Z1) = 1/10$. But the weight of STR2 must be distributed between two state-descriptions, $Z2$ & $Z5$. Carnap chooses to distribute the weight equally within structure-description, so $m^*(Z2) = m^*(Z5) = 1/20$, and in general

$$m^*(Zi) = df \frac{1}{T.z_i}$$

where z_i is the number of Zs isomorphic to Zi. The weight of a proposition will of course be the sum of the weights of the structure descriptions in which it is true.

Now let us return to the inductive problem we ran for $c^\dagger$ and see if c^* does any better. Our hypothesis, h, is Mb & Nb. The

91

null-confirmation is

$$c^*(h, t) = \frac{m^*(t \& h)}{m^*(t)} = m^*(h) = \frac{1}{10.z_1} + \frac{1}{10.z_5} + \frac{1}{10.z_9} + \frac{1}{10.z_{13}}$$

$$= 1/10 + 1/20 + 1/20 + 1/20 = 5/20 = 1/4.$$

So the null-confirmation c^* is equal to $c^\dagger$. But now suppose we again learn *Ma & Na*. Then

$$c^*(h, e) = \frac{m^*(e \& h)}{m^*(e)} = \frac{m^*(Z1)}{m^*(Z1) + m^*(Z2) + m^*(Z3) + m^*(Z4)}$$

$$= \frac{1/10}{1/10 + 1/20 + 1/20 + 1/20} = 1/10 \times 20/5 = 2/5.$$

This means that the function c^* *does* increase in value when the quaesitum property is empirically widespread. Since it also leads to other results of which Carnap approves, he settles on c^* as the proper function for inductive logic.[39]

4   THE SOURCE OF INITIAL PROBABILITIES

Intuition is the principal source of probabilities for Keynes, and logical calculation is the source for Carnap. Keynes allows the Principle of Indifference to be the generator of specific numerical values, while Carnap is more receptive to the use of frequency values as probability values. Otherwise the two systems are quite similar in spirit.

In Dice Games

For Keynes, as for the CTP, the Principle of Indifference is applied directly in cases like dice games and other games of chance. It tells us that the numerical values of the probabilities are equal, and the calculus of probability then enables us to make the usual mathematical predictions. These traditional cases thus receive a traditional treatment based on a revised version of the First Principle of Classical Probability.

In our example of the dice game, Keynes recommends the same procedure as did Laplace: first identify the equipossible alternatives, then assign them equal probabilities and use the calculus of probability to make predictions. Thus, 1, 2, 3, 4, 5, and 6 are the alternatives (here Keynes adds the caution that we must be sure that they are *ultimate* alternatives, i.e., that none of them can be split up into sub-alternatives of the same type). Assigning equal probabilities gives each a value of 1/6. Thus the initial probability of throwing a Five is 1/6. (Keynes further requires that all our evidence be *symmetrical* – that we have no evidence favoring one outcome which is not matched by similar evidence for every other alternative.)

Like the Classical theorists and Lord Keynes, Carnap believes it is possible to assign an a priori value to the probability of rolling a Five, even if neither this nor any other die has ever been rolled. But this is not to be done using the Principle of Indifference, which Carnap rejects because it leads to contradictions. Instead, the probability is derived from the confirmation-function, using the tautology, *t*, as evidence. This initial probability, in the absence of any empirical knowledge, is called the 'null-confirmation of *h*.' But the function c^* always assigns equal null-confirmations to coordinate descriptions,[40] so we still have $P(5) = 1/6$. An added virtue of Carnap's AP theory, however, is that it can also learn from experience. Thus, if we have any empirical evidence at all, such as previous rolls of this or other dice, or physical properties of this die (together with lawlike statements concerning the effects of such properties on dice rolls, to 'connect them up') we can in principle incorporate such evidence in our calculations. (It is currently *practical* to include only information about a small number of rolls of this or other dice.)

Besides the null-confirmation and the probability based on general evidence, a third type of initial AP probability is based on frequencies as evidence. Carnap calls these 'inductive inferences,' but 'statistical inferences' is probably a more descriptive and familiar name. If we know the relative frequency of a property in a sample and seek the probability that *a member of the sample* has that property, we are engaged in *direct inductive inference*. In the present case, though, we desire a *predictive inductive inference*, since the event in question (the next roll) is *not* included in the sample (past rolls). Carnap's rule for predictive inference includes consideration of both logical and empirical factors but is a priori in the sense that any result of a mathematically correct calculation is valid and does not depend on the state of the universe for its truth or its verification. We will discuss this rule and its application when we consider in the next section the probability that a 30-year-old man will get married.

The final source of initial probabilities in a dice game is the relative frequency of Fives in throws of this die considered as a probability$_2$ (RF probability). Carnap thinks that we are *sometimes* justified in identifying the probability of a Five with (the limit of) this frequency, even though we can never know (with certainty) what that frequency is. He, himself, has criticized RF theory quite effectively, but he still thinks it is all right in its place. Unfortunately, *LFP* has little to say about what that place *is*. The most he says for RF probabilities is that our lack of knowledge of frequency limits is not a decisive criticism, since we don't require certainty of scientific hypotheses.[41] Whether this defense is adequate will be discussed in the chapter on RF theories. For now we merely note that Carnap does think the RF theory can generate initial probabilities for dice games in some circumstances, while the various methods of AP probability can generate such values in any circumstances.

Actuarial Cases

While Keynes has rightly been perceived as one of the leading opponents of the RF school of probability theory, it is false to assume, as modern writers on probability sometimes do, that he therefore rejected any use of frequencies in probabilistic arguments.

Consider, for example, the many actuarial cases where frequencies are centrally important. As a practicing economist, Keynes clearly cannot dismiss such cases as uninformative and worthless. On the contrary, he says that 'it is undoubtedly the case that many valuable judgments in probability are partly based on a knowledge of statistical frequencies, and that many more can be held with some plausibility, to be indirectly derived from them....'[42] But such judgments are generally not based on the Principle of Indifference – how then can they be justified?

Keynes gives us one clear answer and hints at a more obscure one. The more obscure answer is that our knowledge of frequencies is part of our knowledge of the world and must therefore be considered in making certain probability judgments. The idea seems to be that my perception of the (non-numerical) probability that the next car I see will be a Cadillac is based on (or at least influenced by) the frequency with which I have seen Cadillacs in the past. To be more precise, the frequency with which I have seen Cadillacs in the past is part of my evidence *h*, and what I directly perceive is the probability relation *a/h* between the evidence and the proposition *a*, that the next car I see will be a Cadillac. The probability in this example cannot be numerical in any theory, since the relevant frequency is not given numerically, but even if it were known to be precisely 0.05, the probability could not be *identified* with the frequency, and might not even be numerically equal to it. There might, for example, be other facts in *h*, such as the fact that I am in a certain neighborhood, which would tend to increase or diminish the probability. Indeed, it is *always* possible that the probability may not equal the frequency, since 'an event may possess more than one frequency, and...we must decide which of these to prefer on extraneous grounds.'[43] (In this case, possible reference frequencies include the frequency with which I have seen Cadillacs in the past year, or ten years, or in this State, or the frequency with which Cadillacs have been produced, or sold, or sold in this State, or....) The best we can do by this obscure method is to let frequencies exert an 'appropriate' influence on our probability judgments.

In the cases where good statistical information is available, however, Keynes gives us a clearer method of statistical induction. This method relies heavily on his theory of induction. To discuss that theory in detail would take us too far afield, but its principal ideas[44] are:

1 Induction is based on probability, which is fundamental.
2 Good inductions depend on Analogy – especially increased negative analogy – rather than simple enumeration.
3 No induction can increase the probability of a proposition unless it has some antecedent (a priori) probability.
4 The sources of these antecedent probabilities are Analogy, the presumption that what we take to be true has thereby some probability of being true, and the Principle of Limited Independent Variety.[45]

By using induction correctly we can make it very probable that certain Laws of Nature (Generalizations) hold in all cases. We can also make it very probable that certain statistical associations (Correlations) found in past experience will continue to hold in the future. We cannot just *assume* that all frequencies will continue at their present value (as Reichenbach does, for example), but in many cases we have sound inductive reasons for expecting them to.[46]

For example, suppose we are trying to establish (*S*): The chance of a male birth is *m*.[47] Assume that data have been collected from all over the world, showing that the overall frequency of male births is *m*. This is not sufficient to establish (*S*) for England, since *m* might be no more than the average value of birthrates that vary widely around the world. (A world-wide frequency of death among infants, for example, would be very misleading if applied to England.) In order to have a strong inductive case for (*S*), we must try to establish that variations in time and place are *irrelevant*. To do this, we break the data down into sub-series which give frequencies of birth in different times and places. If these values do not diverge significantly from *m*, we have much better justification for expecting (*S*) to hold in all cases.[48]

The conclusion of an inductive argument such as this is that the probability – *a/h* – of the proposition, *a*: Male births will occur in the frequency *m* – is relatively high on the basis of all available evidence – *h* – including the fact that such births have occurred in the past in the frequency *m*.

Since Carnap accepts the RF theory as a legitimate form of probability, it is much simpler to just treat actuarial cases as problems in RF probability. Furthermore, as the frequency data become more extensive, the AP method of Predictive Inductive Inference converges to the observed value of the frequency as a limit. So although it is always possible to *interpret* probabilities in such cases as AP probabilities, it is much simpler and just as effective for Carnap to *derive* them by equating them to the observed relative frequency.

5   PROBABILITY OF SINGLE EVENTS

One of the great advantages of the AP theory over the Classical theory and over such frequentist theories as von Mises' is that it can give a theoretical justification to our ordinary practice of estimating the probability of a unique event, in the absence of equipossible alternatives.

If Mr Smith is running for mayor again, and we still wish to know his probability of success, Keynes assures us that we are asking a meaningful question which can usually be answered. All we have to do, according to Keynes, is to assess all the relevant information (judgments of relevance are of great importance to Keynes – they are a major reason why no universal algorithm for determining initial probabilities is possible), and then *perceive* the probability-relation a/h between that evidence h and the proposition, a, that Smith will win.

Now we will ordinarily not be able to intuit a numerical probability in this case. Even if we know the number of possible outcomes (= the number of candidates), it is most unlikely that our evidence for each and all of them will be equal and symmetrical. Therefore, under Keynes's restrictions, we may not apply the Principle of Indifference, and consequently precise numerical values cannot be obtained. What we may be able to do, however, is to make a *comparative* judgment such as 'a/h is more than 1/2 but less than 3/4.' In some cases our intuitions will be strong and precise enough to make this a fairly narrow range, but in many other cases we can do little better than say there is 'some probability' that Smith will win.

Carnap is willing to go much further in this case. His and similar systems are the only type of probability theories which claim that every event has a unique, interpersonal, numerically definite probability on given evidence which is (almost) always theoretically obtainable by calculation. Only they are audacious enough to claim that the probability that Smith will be elected mayor (based on what we know) is a potentially calculable real number between 0 and 1. To obtain this real number by the principal method of Carnap's inductive logic, we must first analyze the structure of a language powerful enough to describe all relevant evidence. Then we take the sum of the m^* values for the state-descriptions in which the evidence is true *and* Smith is elected mayor, and divide it by the sum of the measures of all state-descriptions in which the evidence is true. This gives us the confirmation on the given evidence of the hypothesis that Smith will win.

This method can require an enormous amount of calculation. For the purposes of our example, we will limit the universe to three people: Smith ('s'), Brown ('b'), and Jones ('j'), and only three properties, those of being: Radical (denoted by the predicate 'R'), a Vietnam veteran ('V'), and Mayor ('M').[49] So our language is L_3^3. At this level, we have already 512 possible state-descriptions. (Increasing either parameter by 1 would put us over $z = 4000$.)

Assume that we know Smith is a radical Vietnam veteran and we

97

are interested in his chances of becoming mayor. If we start our list of state-descriptions like this:

$$Z1 = Rs \ \& \ Vs \ \& \ Ms \ \& \ Rb \ \& \ Vb \ \& \ Mb \ \& \ Rj \ \& \ Vj \ \& \ Mj$$
$$Z2 = Rs \ \& \ Vs \ \& \ Ms \ \& \ Rb \ \& \ Vb \ \& \ Mb \ \& \ Rj \ \& \ Vj \ \& \ -Mj$$
$$Z3 = Rs \ \& \ Vs \ \& \ Ms \ \& \ Rb \ \& \ Vb \ \& \ Mb \ \& \ Rj \ \& \ -Vj \ \& \ Mj$$

ringing changes from right to left, we need only run through the first 128 Zs to list all those in which $e(= \text{'}Rs \ \& \ Vs\text{'})$ is true. (See, it's getting simpler already!) Next, we must assign weights to these state-descriptions. $Z1$ is obviously unique, since any rearrangement of the individuals results again in $Z1$ after lexical re-ordering. For $Z2$ through $Z8$, 's' and 'b' have identical properties, so only by substituting for 'j' can we make significant changes. This means that $z_{2,3,...,8} = 3$ (remember that z_i is the number of state-descriptions isomorphic to Zi). But in the most common case each individual will have a different set of properties and there will be 6 ways in which the individuals can be rearranged into different state-descriptions having the same structure. Therefore the most common z_i is 6.

Now we must laboriously sum up the measures for each state-description

$$m^*(e) = \sum_{i=1}^{128} \frac{1}{T.z_i}.$$

Fortunately, a table[50] of Carnap's tells us that the number of structure-descriptions, T, in L_3^3 is 120, so we don't have to enumerate them. Regularities in the table of state-descriptions also simplify the computation, so it is not too difficult to arrive at a value of $180/720$ for $m^*(e)$.

Since we have cleverly arranged our table of state-descriptions to change from the right, we can know that only in the first 64 Zs will '$Rs \ \& \ Vs \ \& \ Ms$' be true. The somewhat easier calculation of $m^*(e.h)$ gives $90/720$.

Finally,

$$c^* (h, e) = \frac{m^*(e \ \& \ h)}{m^*(e)} = \frac{90/720}{180/720} = \frac{1}{2}.$$

But this value, $1/2$, is exactly the same as the initial or null probability that 's' is 'M.'[51] So it seems that our knowledge that Smith is a radical Vietnam veteran has not at all affected the probability of his election. Does that strike you as a bit counter-intuitive? If so, you are no doubt being influenced by the fact that such properties *are* relevant to one's chances of being elected mayor. But *this fact* is not contained in our evidence – we know nothing of who is and is not elected mayor and what their properties might be. We can't learn from experience about mayors until we have at least some experience with mayors.

Suppose we continue our investigation and find that the next city over elected a radical Vietnam veteran named Jones as mayor last year. This adds 'Rj & Vj & Mj' to our evidence, which we will now call 'e & j.' Its range, $R(e$ & $j)$, includes $Z_{1,9,17,25,...,121}$ while $R(e$ & j & $h)$ takes only the first half of these. If we were still using $m\dagger$, and counting each state-description equally, we would again get $1/2$ as our probability. But m^* gives greater weight to those Zs exhibiting homogeneity, so the Zs in $R(e$ & j & $h)$ which attribute similarity to 's' and 'j' count more than those other Zs which say they are different. As a result,

$$c^*(h, e \text{ \& } j) = \frac{m^*(e \text{ \& } j \text{ \& } h)}{m^*(e \text{ \& } j)} = \frac{20/720}{30/720} = \frac{2}{3}.$$

This shows that the odds that Smith will be elected improve if we know that someone with similar properties has already been elected. Such similarities influence our thinking so frequently and so strongly that Carnap has given us another method, the Inference by Analogy, to quickly compute the confirmation gained when they occur. Before we develop that method, we must introduce two more technical terms.

A Q-predicate ($Q1, Q2, Q3, ...$) is an abbreviation for an expression which is 'either the conjunctive predicate expression containing all [predicates] in their alphabetical order ('$P_1, P_2' ... 'P_N$') or is formed from this expression by replacing some of the [predicates] with their negations.'[52]

Our system $L^3 3$ has the following Q-predicates:

$Q1 = R$ & V & M	$Q5 = -R$ & V & M
$Q2 = R$ & V & $-M$	$Q6 = -R$ & V & $-M$
$Q3 = R$ & $-V$ & M	$Q7 = -R$ & $-V$ & M
$Q4 = R$ & $-V$ & $-M$	$Q8 = -R$ & $-V$ & $-M$.

A Q-predicate is the most complete possible description of an individual. It tells of him whether or not he has each property expressible in the language.[53]

The very last technical notion we require (Hurrah! Hurrah!) is that of the *width* of a predicate.

Every self-consistent predicate expression is equivalent to the disjunction of some number, w, of Q-predicates. w is the width of that predicate.[54]

In our system, 'R' has width 4, since it is equivalent to $Q1 \lor Q2 \lor Q3 \lor Q4$, 'R & $-V$' has width 2 ($Q3 \lor Q4$), and so on.

Returning now to the Inference by Analogy, let M_1 be the conjunction of all the properties Smith and Jones have in common ($M_1 = R$ & V). Then w_1 is the width of this expression ($w_1 = 2$). Similarly, let M_2 be the conjunction of all properties that Jones is known to have ($M_2 = R$ & V & M). The width of M_2 is w_2 ($= 1$). Then the confirmation of h by $M_1(s)$ ($=$ our previous evidence e) and $M_2(j)$ ($=$ our additional evidence j) is[55]

$$c^*(h, e \ \& \ j) = \frac{w_2 + 1}{w_1 + 1} = \frac{1 + 1}{2 + 1} = \frac{2}{3}.$$

The reader can see that this is a much simpler way to calculate Smith's chances, if the entire evidence is contained in the analogy. Otherwise the longer method of c^* must be used.

Finally, if we have sufficient statistical information about mayoral elections to establish the relative frequency with which 'people like Smith' have been elected mayor, we may be able to use a Predictive Inductive Inference[56] or even a direct RF interpretation of the frequency as the probability. By speaking of Smith's election as a 'unique event,' however, we intend to express our interest in cases where such recorded repetitions have not occurred, so we will pursue these possibilities no further. Instead, we will briefly consider our other example of a single event, the next throw of a die.

We said in the previous section that Carnap's system assigns a probability of 1/6 to Smith's rolling a Five on the next throw of a die if there is no relevant empirical evidence. So far, he agrees with Keynes and the Classical theorists (assuming our relevant evidence is equal and symmetrical with respect to all ultimate alternatives – Keynes). But if there has been even one throw of a die recorded, the probability changes to 1/7 (if that throw was not a Five) or 2/7 (if it was).[57] In general, c^* is so sensitive to empirical evidence that the results of experience quickly outweigh the initial influence of logical ranges. Carnap's AP probability is very empirical indeed in its willingness to base judgments on past experience – but it is still an a priori method because (1) it *will* give a probability value in the absence of any evidence (and especially in the presence of evidence other than frequencies), and (2) any value correctly derived by this method does not depend on and is not refutable by any experience whatever.[58]

Because of this emphasis on past experience and because of the agreement between the null-confirmation 1/6 and the results of past experiments with many dice, we always take the value of 1/6 as the probability of rolling a Five until experience shows that this die differs from the norm. If a series of rolls of this die gives us some empirical evidence of *any* sort, the Requirement of Total Evidence tells us to include that evidence in our calculation. Since c^* is sensitive to past experience, the new evidence will *alter* $P(5)$ *if* the relative frequency has diverged from 1/6. The probability of rolling a Five with a biased die will fairly quickly depart from 1/6 and converge on the observed frequency. This enables Carnap to claim that his system deals satisfactorily with cases where the simple 'fairness' assumptions of the Classical Theory are violated.

6   PROBABILITY OF REPETITIVE KINDS OF EVENTS

The probability that a man will marry in his thirtieth year, which we have been using as one of our examples of repetitive events, is of the type where Carnap would undoubtedly accept an RF probability (his probability$_2$) as satisfactory. It is characterized by voluminous statistical data, showing considerable uniformity in population sub-groups, and exhibiting a relative frequency other than 0 or 1. In such cases, Carnap's inductive methods converge to the observed value of the relative frequency, so it is much simpler to just *identify* the probability with the relative frequency. But since this chapter is concerned with AP rather than RF probability, I propose to ignore probability$_2$ (which Carnap hardly discusses himself) and instead illustrate Carnap's and Keynes's a priori methods of establishing the probability.

If we have a value for the relative frequency of marriage in an observed sample of 30-year-old males, $rf = S_i/S$, then the probability that a 30-year-old male inside the sample will be married in his thirtieth year is called a direct inference and both Carnap and Keynes agree it is s_i/s. But if we are concerned with a particular 30-year-old man *outside* the sample (as we usually are, in such cases), then what we seek is a singular predictive inference which Carnap defines as:

$$c^*(h, e) = \frac{s_i + w_i}{s + k}$$

where w_i is the width of the predicate 'M_i' ('married in 30th year' in this case) and k is the number of Q-predicates in L. In this case, since the problem is very simple, we get the value $(s_i + 1)/(s + 2)$ for the relative frequency. This, of course, is identically equal to Laplace's Rule of Succession, but in more complex cases it is not. Suppose, for example, we wished to know the probability that our 30-year-old was both married and disinherited ('MD') in the same year. The width of MD is again 1, but k is 4. The desired value is thus $(s_i + 1)/(s + 4)$. In general, Carnap's probability for this hypothesis starts with the a priori (or 'null') prediction w/k, based on the logical width of a property, and then is more and more influenced by observations. As the number of observations becomes very large (in comparison to the complexity of the language) the probability approaches the observed relative frequency as a limit.

In this particular example, let us restrict ourselves to a universe of 30-year-olds and the single property of being married (or not). The width, w_1, of the predicate 'M_1' (married) is 1. The number of Q-predicates, k, is 2 (for '$M_1 x$', '$- M_1 x$'). Thus the initial probability of $h('M_1 x')$ is

$$c^*(h, t) = \frac{s_1 + w_1}{s + k} = \frac{0 + 1}{0 + 2} = \frac{1}{2}$$

as is usual for such predicates in simple systems. Now suppose we observe four such men and find one of them is married. Then the relative frequency of marriage is 1/4, but

$$c^*(h, e) = \frac{s_1 + w_1}{s + k} = \frac{1 + 1}{4 + 2} = \frac{2}{6} = \frac{1}{3}.$$

If we continue to observe the same relative frequency in larger samples, say 1,000 out of 4,000, we find

$$c^*(h, e') = \frac{s_1 + w_1}{s + k} = \frac{1,000 + 1}{4,000 + 2} = \frac{1,001}{4,002} \simeq \frac{1}{4}.$$

The value of c^* can be brought as close to the relative frequency as desired by continued observation.

Thus far we may seem to have been cheating a bit, since the singular predictive inference gives the probability of occurrence of a single event (this man's getting married) rather than the probability of a repetitive kind of event (any such man's getting married). In reality, though, we have almost completed the task of establishing the a priori probability of a repetitive kind of event as the estimate of a relative frequency in a population, because Carnap *identifies* these values:[59]

> Let *e* be any (non-L-false) evidence, *M* any molecular property, *b* an individual and *K* any class of individuals not mentioned in *e*, and *h* the hypothesis to the effect that *b* is *M*, then the estimate (probability$_1$-mean) of the relative frequency of *M* in *K* is equal to the probability$_1$ of *h* on *e*.

In our example, the estimate of the relative frequency of marriage among 30-year-olds equals the confirmation of the prediction that an individual 30-year-old will get married, and therefore approaches the observed relative frequency as the sample size increases. (A more general formula, for the probability of a particular frequency in a population of a definite size, appears on p. 568 of *LFP*.)

In the direct inference case, Keynes would agree that the probability that a man taken at random will marry during his thirtieth year is equal to the frequency, *f*, with which similar men have married in the past, *provided that* we have a strong statistical induction, with good negative analogy as well as many enumerated instances.

The basic justification for deriving probabilities from frequencies is that if we know the relative frequency of a property, *M*, in a *finite* series, *S*, is equal to *f*, then *f* *is* the proportion of favorable to unfavorable cases. To speak of a 'random member' of *S* is to say in different words that 'we have no reason to prefer one alternative over another' or 'the evidence is equal and symmetric with respect to the ultimate alternatives.' But these are just the necessary and sufficient conditions for the application of the Principle of Indifference, so that in this case, 'by a straightforward application of the Principle of Indifference we have $p = f$'[60] (where *p* is the probability that a random member of *S* will have the property *M*).

Keynes thus has a sound theoretical account of the simplest type of statistical problem, sampling from a known population. But when we go on to more varied statistical problems we encounter difficulties. Keynes identifies two types of problems:[61]

1 The *statistical problem* of attempting to establish the true relative frequency when we ordinarily are given only selections from the series involved, not the entire series.
2 The *inductive problem* of arguing from the frequencies in the known series to the frequency in the desired series.

It is difficult to see how the Principle of Indifference can carry us far toward a solution of these and other advanced statistical problems, and, indeed, Keynes presses no further. In the example of the marriage, therefore, the most we can say is that good, varied, statistical evidence will give a very high probability that the value of *p* is very close to *f* (always remembering that we are talking about a random man of thirty and not some definite individual).

In the other type of repetitive event, where we are concerned with the probability that a (some, any) throw of a die will be a Five, Keynes reverts to the Classical solution of Indifference. He does, however, caution us that the total evidence must be symmetric, and the frequency of Fives thrown so far by this die is part of that evidence. The probability 1/6 is *not based on* the evidence concerning past frequencies, it is based on the Principle of Indifference. The only reason for checking past performance is to make sure that application of the Principle is not precluded by asymmetrical evidence (evidence of bias).

Indeed, Keynes has nothing but scorn for attempts to base probabilities on empirical, a posteriori observations when a priori methods are available, or attempts to confirm or deny probabilistic conclusions by empirical test. He mentions,[62] as a curiosity, 'the enormously extensive investigations of the Swiss astronomer Wolf' who 'recorded altogether...in the course of his life 280,000 results of tossing individual dice.' Most of the results were close to theoretical expectations, but in one experiment of 20,000 tosses the a priori predictions were violated considerably. Could this be an empirical disproof of some element of probability theory? No, says Keynes,[63] because

The explanation of this is easily found; for the records of the relative frequency of each face show that the dice must have been very irregular, the six face of the white die, for example, falling 38 per cent more often than the four face of the same die. This, then, is the sole conclusion of these immensely laborious experiments, – that Wolf's dice were very ill made.

I invite the reader to reflect seriously on this whimsical example, for it illuminates one of the central features of a priori probability which has perplexed many people in the past – *the theory is never wrong and neither it nor any of its predictions can be refuted by experience.*

The simplest reason for this unassailability, which Keynes appeals to in laughing off Wolf's experiment, is common to Classical theories as well. It is that all statements and predictions concerning games of chance *assume that the game is fair.* They are generally worded, 'For an unbiased die, the probability of...' or 'Given a well-shuffled deck,...' If these qualifying phrases are not explicitly present, they will always be invoked as 'implicit assumptions' or 'basic conditions for the application of the theory' when things go wrong. If the qualifiers are explicit, the entire prediction has the form 'If the dice are fair, then the probability of a five is 1/6.' In conventional symbols:

$$(1) \ p \supset q.$$

Now suppose we run the experiment and the prediction fails, i.e., we discover that $-q$ is true. We now have two alternatives, we can either abandon (1), or we can declare that p is false. The theorist of course prefers to say that p is false. Is that a legitimate move or is he just refusing to accept the recalcitrant evidence?

It depends, obviously, on what we *mean* by 'fair dice.' If this were an empirical theory, 'fair dice' might refer to certain physical features of regularity, uniform density, etc. But for a priori theories in general, as A. J. Ayer says, 'A true coin, or a true die, is simply defined as one that yields results which are in accordance with the a priori calculus of chances. But then all these judgments become mathematical truisms.'[64] It should be clear to the reader that no empirical evidence can ever refute a mathematical truism.[65] It follows, then, that any die which fails to perform as expected is *per definition* 'unfair', and the probability that a (fair) dice throw will yield a Five remains as determined by the Principle of Indifference.

In the problem of predicting the throws of Five with a given die, Carnap's system has the advantage that it can deal with either fair or biased dice. The problem can be treated in the same way as the marriage example, although presumably the initial probability will be somewhat more reliable. This reliability is based on the nature of dice rather than any feature of Carnap's system. Since dice are (normally) designed to exhibit an empirical division in the same proportion as the logical division of predicates describing dice results, we should not be surprised by the regularity with which they do so. Any deviation from this regularity will be detected by c^*, and the probability for the type of event (a Five) will alter just as the probability of a single event did in the previous section, since the two probabilities are, on Carnap's view, identical.

7 PHYSICAL CHANCE AND ABSOLUTE PROBABILITY

Keynes was writing in the cheerfully deterministic time before the ascendancy of quantum mechanics. Neither he nor those he criticizes 'wish to question the determinist character of natural order...,'[66] but all wish to give some sense to the phrase 'objective chance.' Keynes discusses,[67] for example, Cournot's theory that objective chance results from the convergence of independent series of events and Poincaré's theory that objective chance arises when slight variations in the same cause result in great variations in the effect. He notes that such definitions as these mean that the real source of 'objective chance' is our own ignorance or inability to perceive and analyze complex and varied causal chains. This shows, he says,[68] that 'objective chance' is really just a sub-species of 'subjective chance' (= Keynes's own AP probability). In this spirit, he offers the following definition,[69] which is intended to pick out those situations (such as dice games) where we are most inclined to think of some objective chance in operation:

> an event is due to objective chance if in order to predict it, or to prefer it to alternatives, at present equiprobable, with any high degree of probability, it would be necessary to know a great many more facts of existence about it than we actually do know, and if the addition of a wide knowledge of general principles would be little use.

This disposes of objective or physical chance – Keynes does not think that the universe is so constructed that some things happen 'by chance alone.' He is a classical determinist in his physics.

Carnap, of course, was thoroughly familiar with modern challenges to Newtonian mechanism. How does he feel on the question of physical probability? *Is there* a value $P(X)$ which obtains in the real world and measures the chance that an event, X, will occur, or is there no randomness in nature? (A terminological point which concerns Carnap: a series of events or numbers is *random* if the constituents of the series are not related by any laws whereby the value of one determines or influences the value of another and there is no external law or process which determines them:[70] the same series is *disordered* if there are relatively few universal laws which apply to it, or, less stringently, if there is relatively little similarity or functional ordering of the individuals. Thus, 8888 is a highly *ordered* number, but it may yet be a *random number* if produced by a random number generator, while 3.14159 is a comparatively *disordered* number but hardly ever occurs randomly. Carnap contends that many statisticians and scientists confuse disorder with randomness.)[71]

LFP makes no pronouncement on the concept of physical probability. In it, Carnap has nothing to say about determinism or pure chance, so to be true to our intended restriction to that volume, we should perhaps just say that we *don't know* what Carnap thought. But a few remarks might be helpful, none the less.

First, it is clear that there can be no objective or physical probability if 'probability' is understood as Carnap's 'probability$_1$,' since this latter is a property of pairs of sentences rather than a property of empirical objects. It should be equally clear that accepting this terminological restriction has no bearing on whether or not every event in the universe is fully determined by previous events and universal laws.

Second, the universe of *LFP* has been described as 'unchanging'[72] since the individuals cannot be ordered in space-time and since the specification of the true state-description determines the facts once and for all. This might be taken to mean that nothing happens by chance since, indeed, nothing happens. I reject this interpretation for two reasons: (1) Carnap never intended the languages *L* to be adequate for science, and any defects they have should not be interpreted as necessarily applying to the universe (he surely did not intend the restrictions on *L* to mean, e.g., that all properties are independent in real life), (2) whether or not some events *do* occur partly by chance, so that 'red at x, y, z, t_1' probabilifies 'red at x, y, z, t_2' but no description of the former sort, together with general laws, *implies* the second description, an extensionally complete description of the history of the universe need only include both sentences, not any causal or logical connections between them. State-descriptions are intended to tell only what *is* the case, not *why* it is the case. The fact that the true state-description assigns Green to a space-time point in my vicinity does not mean that that point *could not* have been Red, or had no chance of being Red, or was determined or fated to be Green – it means only that that point *is* Green (in the omnitemporal sense of 'is').

Third, in his 'Intellectual autobiography' Carnap describes his early ideal of a system of physics as a Laplacian determinism which allows calculation of a complete description of the universe at any time-point, provided such descriptions are known for two arbitrary time-points.[73] The autobiography does not say whether his later work in quantum mechanics led him to abandon this ideal or whether his friend Einstein influenced him to retain it; all we know is that for at least part of his life, Carnap was a determinist.

Of course the question of physical chance is a metaphysical/cosmo-logical problem, on which a logical theorist need not have an opinion. But its logical or mathematical correlate is the notion of absolute probability. Surprisingly, Keynes and Carnap both are willing to speak (guardedly) of the absolute probability of something.

Ordinarily one would think that 'the absolute probability of X' would stand in contrast to 'the relative probability of X' so that the latter is dependent on some evidence while the former is not. But the inherently relational nature of AP probability makes it *meaningless* to talk about probabilities in this abstract fashion.

> It is as useless, therefore, to say 'b is probable' as it would be to say 'b is equal,' or 'b is greater than,' and as unwarranted to conclude that because a makes b probable, therefore a and c together make b probable, as to argue that because a is less than b, therefore a and c together are less than b.[74]

AP probabilities always vary with the evidence just as distance varies with the starting-point ('How far is China?' 'From where?'). But what happens when one has collected all available evidence? Then, Keynes thinks, one does indeed have *the* probability of X. He does not even require that we include all *possible* evidence, merely all evidence now known to us.

> If h specifies the Universe of Reference, *i.e.* if its group compre-hends the whole of our knowledge, p/h is called *the absolute probability of p*, or (for short) *the probability of p*; and if $p/h = 1$

and *h* specifies any real [i.e., known to be true] group, *p* is said to be *absolutely certain* or (for short) *certain.*[75]

Carnap is in complete agreement with Keynes's dictum that probability is always relative to some evidence. There is likewise no room in his system for an absolute probability independent of any evidence and he identifies the absolute probability of *X* with the probability all things considered.

There is, however, a role – and a very special role indeed – for probabilities which are independent of any *empirical* evidence. These are called 'null-confirmations' by Carnap. In other systems they might be labelled 'initial' or 'a priori' probabilities. One of their functions is to enable Carnap to specify the probability of any given sentence in a language system, based solely on considerations of logical range. This is an accomplishment peculiar to Carnap's system (and others similar) of AP probability, which already makes it more powerful than most. But it has another consequence which might be of even greater practical importance – it guarantees that initial probabilities are always attainable for the application of Bayes's Theorem.

The use of Bayes's Theorem (or the 'inverse' of Bernoulli's Theorem) to convert relative frequencies into probabilities has been a bone of contention among probability theorists throughout the history of the field. In a cogent and detailed argument,[76] Keynes had sought greatly to curtail applications of such formulae on the grounds that one almost never possesses the required numerically definite values for the initial probabilities. Carnap agrees that such prior probabilities are required for the application of the theorem, and further agrees that most other systems of probability are incapable of generating them.[77] But in *his* system , $c^*(h, e)$ is *always* computable (for finite systems), whatever *h* and *e* might be. Thus, it seems that Bayes's Theorem will be applicable in many of those cases where Keynes had disallowed it.[78]

But useful as null-confirmations are, they are not what Carnap would mean by '*the* probability of *X*.' We might be inclined to say that there is *no* meaning for this phrase in Carnap's system – and in one respect this is so. Clearly every hypothesis, *X*, has a vast (perhaps infinite) number of different probabilities based on the many different possible sets of evidence *e, e', e''*, etc. Nothing in the system distinguishes any one value over any other. Therefore we are unable to identify *the* probability of X. But isn't this, after all, what we truly were seeking?[79]

It is true that nothing in the formal system Carnap constructs gives any preference to one of these probabilities over another, but it does not follow that we therefore have no idea which to choose. If I am interested in the probability that it will rain tomorrow (h), and I choose to evaluate it on the basis of evidence (e) that the Premier of China loves eggs, Carnap's system will (ideally) give me a definite numerical value for $c(h, e)$. If I choose to act on this value, I will break no formal rules of inductive logic, but I *will* be acting rather foolishly.

The foolishness of my act arises here from the fact that I am acting on evidence which is irrelevant to my problem and ignoring evidence which is relevant. The concept of Relevance of Evidence is a difficult one – it seems likely that there can be no general rules for determining in advance which evidence is relevant and which is not. The usual 'definition' of Relevance is something like 'A sentence (or event, etc.) A, is relevant to another sentence (etc.), X, if and only if A is *not* stochastically independent of X.'[80] But stochastic independence of A and X means only that $P(X) = P(X|A)$, that is, the probability of X given the fact (or knowledge, or assumption) of A is no different from the original probability of X.[81] This obviously gets us no forrarder, since the whole point of asking about the relevance of A is to decide whether or nor we *should* include it in our calculation of $P(X)$. If deciding the relevance requires us to calculate $P(X)$ and $P(X|A)$ and see if they are equal, it would have been simpler to just go ahead and include A in the evidence to begin with. If A *is* stochastically independent of X it will have no effect on the calculated value of $P(X)$; if it is *not* independent (and therefore it *is* relevant) we must be careful not to leave it out. Therefore we should include A in our evidence. But the same argument applies to *every* sentence which is known to be true but not known to be stochastically independent of X. Does that mean we must include in the evidence everything that we know?

That is precisely what it means, according to Carnap. To ensure against errors of omission when we calculate $P(X)$, we must take our total knowledge as evidence. This is what he calls the

Requirement of total evidence: in the application of inductive logic to a given knowledge situation, the total evidence available must be taken as basis for determining the degree of confirmation.[82]

This requirement (which he attributes to Bernoulli, Keynes, and Peirce, and finds sadly lacking in Laplace) is 'not a rule of inductive logic but of the methodology of induction...'[83] Thus we select one of the many possible values of $P(X)$ not according to some principle in a formal system but in deference to a rule which guarantees the inclusion of every relevant sentence which is known. The result is what Carnap calls '*the* probability of X' (although this is not a prominent phrase in Carnap – he prefers to stress the relativity of probability to given evidence, e, and then let e be everything we know).

But Ayer's argument can be pushed one step further. Even if we agree that the requirement of total evidence is a reasonable guarantee that we will achieve *the* probability of X, relative to everything we know, this is still no reason for thinking that it is or even approximates *the* probability of X *in the real world*, and that, again, is the real point of the game. The aim of probability theory is not to tell us things about logic and propositions and evidence but to enable us to predict what will probably happen in the real world.

Carnap does not specifically answer this objection, but I think we can construct a reasonable reply from his general attitude and by analogy to the remarks he made about abstraction.[84]

It is a fundamental principle of rationality (at least for empiricists) that one should base one's actions and judgments on 'the facts'. But we must not be misled by the ambiguity of this latter phrase. It is true that it sometimes has the *ontological* meaning 'the facts as they really are,' which refers to the true (description of the) state of the universe. It is these facts which ultimately determine the success or failure of our actions. If we 'acted on' these facts, we could indeed maximize our chance of success. But how is one to 'act upon' the true state of the universe? Empiricists generally agree that most, if not all, of our empirical knowledge is at best probable – we have no certain grasp of the true state of the universe. What we do have is knowledge of 'the facts' in the *epistemological* sense of 'the facts as we know them,' which consists of those of our empirical beliefs which are most highly warranted. We can 'act upon' these 'facts,' since they are known to us and can enter into our deliberations. But acting on *these* facts may or may not maximize our chances of success. That depends ultimately upon how much congruence there is between 'the facts as we know them' and 'the facts as they really are.' It is one of the chief difficulties of the human condition that 'the facts' which would save us are unattainable and 'the facts' which we can acquire give no guarantee of success.

It should be obvious by now that this is *not* a difficulty peculiar to Carnap's probability theory. The answer to the question 'Why should we act on a probability based on our evidence only, which may or may not correspond to the value in the real world?' is that we are *always* constrained to act on our evidence only and *hope* that it corresponds to the real world. It would be irrational to do otherwise. Even if one *could* give good metaphysical sense to '*the* probability of *X*,' all that we could ever act upon would be 'the value we think, based on what we know, is probably *the* probability of *X*'. The RF school and Ayer tend to ignore this unfortunate fact in their search for the objective probability of *X*.

8   THE METAPHYSICAL STATUS OF P

If *P* is a probability statement in an a priori system, it is held to have no metaphysical import at all. It is 'not a question of facts'[85] but a purely formal statement which, if true, is *L*-true (logically true; the '*L*' does not refer to the languages *L*).[86] It is 'independent of the contingency of facts because it does not say anything about facts.'[87] We are to construe it like '(*A* & *B*) ⊃ *A*', which is true and can be known to be so whatever the state of the universe might be.

Here Carnap is in complete agreement with Lord Keynes – indeed this attitude can be considered the defining feature of AP (or 'logical') theories of probability. The essential point is that a probability determination is a logical relation between sentences (or, for Keynes, propositions) and says nothing about the world. If it is true, it is logically true and can never be falsified.

For Keynes there is, however, one metaphysical presupposition behind a probability statement, *P*, or, more precisely, there is a metaphysical ground of *P*'s validity. This ground or basis of *P*'s truth is the existence, in its present form, of the human mind. Keynes explicitly denies that probability-relations are objective in the sense that any rational being would agree to them. Instead he says that they are 'the degree of probability to which those logical processes lead, of which our minds are capable....'[88] There would be no point in trying to escape this restriction because

> If we do not take this view of probability, if we do not limit it in this way and make it, to this extent, relative to human powers, we are altogether adrift in the unknown; for we cannot ever know what degree of probability would be justified by the perception of logical relations which we are, and must always be, incapable of comprehending.[89]

The validity of *P* depends, therefore, in a very Kantian sense, on the constitution of the human mind. We might even press the analogy further by saying that the validity of individual probability-relations depends upon the nature of our perception (cf. Kant's 'Forms of Intuitions'), while the validity of the probability calculus depends upon the nature of our reason (cf. the 'Categories of the Understanding').

Now, we might ask ourselves, since Keynes's probability-relation is based on the nature of human beings, like Kant's famous principles, is it also, like those principles, supposed to be regulative of all possible human experience? That is to say, can we know a priori that all possible experience will conform to the probability calculus and our perceptions of probability? This is a complex and difficult question which Keynes doesn't address directly. It may help if we distinguish some different senses in which the question might be intended.

First, if we intend to ask whether it is a metaphysical necessity that *a* is true whenever *a/h* is very high the answer is 'no.'[90] This is so because of the generally accepted fact that even a very probable event need not occur,[91] and also because the probability of *a* is only asserted relative to the evidence *h*, and if some other factor, *h'* is involved, $a/h \cdot h'$ might be very small.

If we ask whether *a* must always occur when $a/h = 1$ and *h* is known to be true (a tautology, perhaps) Keynes's answer seems to be 'yes.' At least he says of such a proposition that it is certain.[92] Now 'certain' is, by standard usage, an epistemological term, whose metaphysical correlate (usually) is 'necessary.' Keynes does not assert, but we may reasonably attribute to him, the common view that the occurrence of any event described by a certain proposition is necessary. If Keynes would accept this principle, then his answer to this question would clearly be 'yes.'

Again, if we ask whether the world is such that the entire

proposition '$a/h = p$' is true whenever we are warranted in believing it, the answer is 'yes.' In this case, however, it is *trivially* 'yes.' Since a probability statement says nothing about the world, it is compatible with any metaphysical reality whatsoever.

Finally, if we ask whether the world is such that '$a/h = p$' might ever be falsified *when applied to the world*, the answer is indeterminate. It is, indeed, precisely parallel to asking whether '$2 + 2 = 4$' can ever fail in practice, and this is clearly not a question to which a general answer can be given. If the interpretation is 'When 2 things are put with 2 other things you have 4 things together in one place' then it can indeed fail in experience. For miscible liquids, 2 quarts of this put with 2 quarts of that result in somewhat less than 4 quarts altogether. If every two objects in the universe were 'miscible' with every other object in a fixed ratio, it seems quite reasonable to think that applied or practical arithmetic might be quite different. (I leave it to the reader's intuitions to decide whether such a universe 'fails to obey the laws of addition,' or, even worse, 'falsifies' them.) Keynes thinks that the utility but not the validity of both addition and probability theory depend upon the state of the universe. Thus,[93]

> If there were no repetition of detail in the universe, induction would have no utility. If there were only a single object in the universe, the laws of addition would have no utility. But the processes of induction and addition would remain reasonable.

It seems, then, to be a contingent metaphysical fact that the universe obeys the principles of probability theory. There is only one correct system of probability (in whatever sense of 'is' makes sense). Its pragmatic value, but not its truth, depends upon the state of the universe.

Carnap nearly agrees with this in *LFP*, but in his later writings (especially *The Continuum of Inductive Methods*) he tends to abandon the single function $c*$ in favor of a continuum of possible inductive methods, characterized by the parameter λ (which can be taken as a measure of 'degree of caution in drawing inferences'). Thus we have the task of choosing our inductive method before (during? after?) our attack on the problem. On what grounds do we make this choice?

In *LFP*, when Carnap was still defending $c*$ as the only proper confirmation-function, he set about it this way:[94]

We shall discuss various definitions for concepts of degree of confirmation or requirements for such definitions. We shall see that it is often hardly possible to judge the plausibility of definitions or requirements, that is, their adequacy for an explication of probability$_1$, by merely inspecting the definitions themselves. The judgement is rather to be based on an investigation of the consequences to which the definitions or requirements lead. Thus, the plausibility of a definition is judged by the plausibility of the theorems derived from it; and this in turn can often be judged in the easiest way by studying concrete numerical examples.

The idea of judging *c*-functions by their consequences might well lead one to think that Carnap adopts a pragmatic attitude towards the assessment of probability systems. This is not quite the case. When he sets out to evaluate the *consequences* of the system, he doesn't ask whether they lead to success in practice, as a pragmatist would; nor does he ask if these consequences are in agreement with reality, as a correspondence theorist might. Instead, he asks whether they correspond to *our intuitive judgments* on the matter. Applying this same method to the choice between whole systems of inductive logic, Carnap later writes:[95]

It seems to me that the reasons to be given for accepting any axiom of inductive logic have the following characteristic features... :
(11) (a) The reasons are based upon our intuitive judgments concerning inductive validity, i.e., concerning inductive rationality of practical decisions (e.g., about bets).
Therefore:
(b) It is impossible to give a purely deductive justification of induction.
(c) The reasons are a priori.

Carnap's method of justifying (or even just choosing) a system of probability seems to lead us ultimately back to the same bedrock as Keynes's did – human beings *have* intuitions about probabilities and any satisfactory system for the formalization of probability must conform to those intuitions.

This result could have been anticipated, perhaps, by reflecting on

Carnap's *aim* in *LFP*. It is his intention, he says, to give an *explication* of terms like 'degree of confirmation' and 'probability.'[96]

Since the primary requirement for an explication is similarity to the term explicated,[97] it follows that the primary requirement of a theory of (AP) probability is that it conform to the existing ideas, practices, and intuitions of human beings in matters of probability. The fact that such a theory *also* conforms to the external world (by leading, e.g., to success in betting or in estimating empirical frequencies) is surprising only if we fail to reflect that (1) the world to which it conforms is the world as perceived by that same human mind, not (necessarily) the noumenal world, and (2) those ideas, practices, and intuitions were themselves evolved (God-given, developed) for the specific purpose of successfully dealing with the world in just such situations. So far as the human mind successfully represents reality, and so far as a theory of probability successfully represents human intuitions and judgments, just so far will the theory conform to reality. But the *primary* aim of AP theories of probability is to conform to the mind and not the world.[98]

9   THE EPISTEMOLOGICAL STATUS OF P

Our probable knowledge, like the rest of our knowledge, is fundamentally based on intuition, according to Keynes's epistemology.

All our knowledge is divided by Keynes into that which is direct and that which comes from argument.[99] There are three sources of direct knowledge (Keynes seems to prefer 'direct knowledge' and 'direct acquaintance' to 'intuition'):[100]

1 We *experience* our own *sensations*.
2 We *understand ideas* and *meanings*.
3 We *perceive* '*facts* or *characteristics* or *relations* of *sense-data* or *meanings*.'

One of the most important types of things with which we are directly acquainted through perception is the logical relation between two propositions. If we have direct knowledge of *R*, it may happen that we will also have direct knowledge of '*R* ⊃ *S*', in which case we will have 'indirect knowledge,' or 'knowledge by argument' (rather than by acquaintance) of *S*. But *modus ponens* is not the most basic of these relations – that honor goes to the probability relation $a/h = p$. For Keynes, 'the laws of inference are the laws of probability,... the

116

former is a particular case of the latter.'[101] Here, for example, *modus ponens* is just the special case where $a/h = 1$.

Some among these relations of probability are so general and so fundamental that they constitute a body of axioms and theorems which correspond to the traditional Laws of Thought, but go beyond those Laws 'in dealing at the same time with the laws of probable, as well as of necessary, inference.'[102] Chapter XII is devoted to setting forth this select body, which is capable of generating all formal truths of inference. All of these relations are known through direct acquaintance and suffice for formal logic and probability theory, but if we wish to include Induction as a branch of probable reasoning, we require one more principle, with a somewhat different epistemological basis.

The required principle, which Keynes called the Inductive Hypothesis, has come to be known as the Principle of Limited Independent Variety.[103] It asserts that, in the relevant universe of discourse, all differences between objects arise from a finite number of generator properties, or, to put it another way, 'that the amount of variety in the universe is limited in such a way that there is no one object so complex that its qualities fall into an infinite number of independent groups....'[104] This principle assures us that *every* empirical generalisation, or statistical correlation, has at least *some* finite probability *a priori*. Once this initial probability is established, we can use what is basically a variation on Bayes's Theorem, together with consideration of analogy, to increase the probability through experience.

The Principle of Limited Independent Variety is first introduced as a logically necessary presupposition for the validity of induction, but Keynes then goes on to argue that it is a justifiable assumption on grounds that closely resemble Kant's notion of synthetic a priori truths. It is known, he says, in the same way that we know the Principle of the Uniformity of Nature and the Law of Causation.[105] Such knowledge is *synthetic* because it goes beyond 'what may be regarded as an expression or description of the meaning or sensation apprehended by us.'[106] The principles are not strictly a priori because we cannot know in advance 'that [they] would be equally applicable to all possible objects.'[107] But when we have at least some experience of phenomenal objects, 'we have a direct assurance that in their case...the assumption is legitimate. We are capable, that is to say, of direct synthetic knowledge about the nature of the objects of our experience.'[108]

In the end it seems that, while Kant's synthetic a priori knowledge derives from reflecting on and thinking about our forms of knowledge and intuition, the principles that Keynes accepts are just a *product of* one of our forms of knowledge. We have the faculty to know such non-inductive generalizations about experience, and there's an end on't.

Returning now to lower level probability judgments, we find that some are derived through the probability calculus and are therefore known by argument; but of course the probability calculus generates no probabilities unless at least some probabilities are available as data, and these *initial* probabilities must be known directly. Many philosophers have balked at Keynes's assumption that we somehow have the faculty of directly perceiving probability relations. Keynes himself did not seem particularly concerned to identify or analyze the faculty, or justify his belief in it. Hilary Putnam has suggested that Keynes's lack of concern over assuming such a faculty is a direct result of his earlier deep acceptance of G. E. Moore's ethical theories.[109] Whatever the reason, it just *is* the case that Keynes assumes throughout his work and never tries to explain or justify an innate human ability to perceive the probability relation between propositions.

Not even Keynes, however, was extreme enough in his intuitionism to think that we could always and easily determine an exact numerical probability for any proposition on any body of evidence. Instead, he asserted his famous doctrine that some probabilities are non-numerical and some non-comparative. Perhaps the best way to substantiate this claim is to examine our own intuitions. I think, for example, that there is some probability that the Dean will resign. But I can certainly not specify a numerical probability value for that event. And if I am asked whether I think it more likely that the Dean will resign or that the new library will open on time, I will be unable to answer.[110]

There is, however, one class of cases in which we *do* perceive definite numerical probabilities. This is the class of cases where the Principle of Indifference can be applied. Besides renaming the Principle of Insufficient Reason, Keynes also revised and restricted it. He sought to avoid paradoxes by requiring the alternatives to be ultimate rather than divisible, and he sought to avoid mistakes by requiring that all relevant evidence be symmetrical with respect to the alternatives. In these – and only in these – cases we can confidently apply the Principle of Indifference to arrive at precise knowledge of the equiprobability of the alternatives.

Again the epistemologist who wishes a justification of principles will be disappointed. Keynes makes no effort to explain why symmetrically-evident alternatives must occur with equal frequency or even to explain why we must *think* that they do so. He seems to believe it self-evident that *some* version of the Principle of Indifference is essential to probability theory. His only concern is to find a version which is both useful and reliable, and it is to this that he devotes his attention.

But even with the restrictions and refinements Keynes added to the Principle, it is not sufficient *by itself* to generate the initial probabilities without depending on direct perception. Nor will any other rule be able to do this: 'there is little likelihood of our discovering a method of recognising particular probabilities, without any assistance whatever from intuition or direct judgement.'[111] The reason for this is that we must rely upon direct judgment to tell us when the Principle is *applicable.* For example, in a dice game we know that the six faces of the die are different – they are located in different places and they have different patterns of dots painted on them. Thus we have asymmetrical evidence for each side and the Principle may not be applied – unless we can eliminate the differences as irrelevant. But we have no rule for identifying irrelevant considerations, so we must make a direct judgment in this case. But in *every* case there is *some* difference between the alternatives (cf. Leibniz's Principle of the Identity of Indiscernibles), so it follows that in every case we must make some judgment of relevance in deciding whether to apply the Principle.[112]

When all of these conditions are fulfilled, and we apply the Principle of Indifference, it serves as an aid or tool for our intuition, enabling us directly to perceive the proposition '$(a/h) = x$.' Since this knowledge is direct, it is also certain,[113] but it is generally only knowledge *about* a and not knowledge *of* a. Only in the case where $x = 1$ can we be said to have knowledge *of* a, but when we manage this, our knowledge is certain.[114] An obvious precondition of the certainty of a is that the evidence itself, h, must also be known. Indeed, Keynes imposes this restriction on *all* probability implications.[115] It is allowed that h may contain some 'secondary propositions' – statements of probability-relations, like '$a/h = p$' – but it is required that these and all other members of h be *known.* To be

119

known, such statements must be true and not just probable. It follows, then, that all our probable knowledge is based on some certainty. Keynes's system does not allow us to conclude anything from a proposition if our degree of rational belief in it is less than certainty.[116]

> I assume then that only true propositions can be known, that the term 'probable knowledge' ought to be replaced by the term 'probable degree of rational belief,' and that a probable degree of rational belief cannot arise directly but only as the result of an argument, out of the knowledge, that is to say, of a secondary proposition asserting some logical probability-relation in which the object of the belief stands to some known pro-position. With arguments, if they exist, the *ultimate* premisses of which are known in some other manner than that described above, such as might be called 'probable knowledge,' my theory is not adequate to deal without modification.

The spirit of all this is very similar to C. I. Lewis's epistemology, and I think it likely that if Keynes had constructed a complete epistemological system, its similarities to Lewis's would have been far more pronounced than its differences.

Carnap's epistemological views, on the other hand, are much more like the Logical Positivists'.

If P is a statement within Carnap's system of probability which has the form $c^*(h, e) = p$ (for a given language, L) the question of P's truth (and likewise our knowledge of it) is a purely *semantic* question. The c-function is *effective* in the sense that if we completely grasp the structure of L, and the meanings of its terms (and the rules of deductive logic and the c-function) we can always decide whether or not P is true by strictly a priori means (provided the system is finite). This means that P is 'L-determinate' (i.e., its truth or falsity is a logical question) and, if true, it is L-true.

It follows from all of this that any c^*-statement will be agreed to by all competent investigators who take the trouble to investigate it – a considerable epistemological virtue.

But there are agreements and then there are agreements. In particular, I can agree that P, but *deny that the probability of h is p*. (This point is clearly and exhaustively analyzed in A. W. Burks's essay 'On the significance of Carnap's system of inductive logic for the philosophy of induction,'[117] from which the following argument borrows heavily.)

If I believe that the probability of *h is p*, I will normally be willing to act on that belief. For example, I will be willing to bet on *h* at any odds equal to or better than $p:(1 - p)$.

But if I accept a different measure function, say $m^\dagger$, or adopt some other theory of probability entirely, I may argue that the probability of *h* is *really q* – even though I agree that *P* is true a priori (logico-mathematically).

So a particular proposition, *P*, may be well-founded and completely warranted within Carnap's system, but its acceptance as a fact about the world will still depend on our epistemic attitude towards the system as a whole. Here Carnap argues that:

1 We clearly do have a concept of logical probability.
2 Although imperfect, $c*$ is the best explication of that concept yet developed.

If we accept these premises, Carnap thinks, then we should agree that all $c*$-statements are sufficiently warranted until and unless someone develops a *better* system of quantitative inductive logic.

The epistemological status of *P* is therefore analogous to that of a theorem in a system of geometry – it is logically true within the system, but one must present evidence to justify the conclusion that it applies to reality. The difference between the two is that the 'reality' to which a geometrical statement must apply is physical, while a probability statement is intended by Carnap to apply to our inductive intuitions and behavior.

10   THE RATIONALITY OF PROBABILITY BEHAVIOR

Keynes is concerned with delimiting and organizing the *kinds* of probability behavior which are rational, but he evidently sees no possibility at all that the entire enterprise might be misguided. On the very first page of Chapter I he says that 'in the actual exercise of reason we do not wait on certainty, or deem it irrational to depend on a doubtful argument.'[118] He continues this attitude throughout the book, asking always which beliefs are rational and which fallacious, but never seeking to explain the nature of rationality in general.

Keynes's attitude is partly explained by his epistemology. Our knowledge of the rules of probability is direct knowledge and it is epistemically prior to the rules of deductive logic. If we take deductive logic as the paradigm of rationality (as people were still doing in Keynes's day) it would be illegitimate to use that logic against its own foundations. Therefore the rationality of probability theory is beyond question – it is more fundamental than our method of questioning it.

As noted above,[119] there is one passage in which Keynes admits that the *utility* of probability behavior depends on the nature of the universe, but he denies that its rationality does! It follows from this that probability theory would be rational even if it were not useful. Clearly Keynes is using the word 'rational' to mean something much more like 'in accordance with the Laws of Thought' than like 'leading to success in practice.' But on this definition it is *analytic* that probability theory is rational, since the Laws of Thought *are* the laws of probability theory.

The reader will note that I have equivocated by starting out to talk about probability behavior and ending up talking about probability theory. The former is my concern, but Keynes deals only with the latter. In his work on probability, Keynes is far more the son of his father the logician than of his mother, the Mayor of Cambridge. He is concerned with theory, not action. Perhaps the best interpolation of his views on action would be that in *this* world it is (contingently) wiser and more conducive to success to act on rational beliefs, and probability theory tells us which beliefs are rational.

We have examined already Carnap's argument that c^*-statements are warranted because they are the best available probability statements. But do we really *need* probability statements – can Carnap show that our probability behavior is rational or, what might be the same, that it will lead to success in practice?

Obviously he can't show that any *one* action taken on the basis of probability will succeed, since the very concept itself involves some possibility of failure. But it would be sufficient if he could show that *in the long run, X* will be successful if he acts on probability.[120]

> If X could know this, then he would clearly be justified in following the inductive method. It is clear that the truth of [this hypothesis of success] is not logically necessary but depends on

the contingency of facts. Statements like [this] which assert
success in the long run for the inductive method would be true
if the world as a whole had a certain character of uniformity to
the effect, roughly speaking, that a kind of events which have
occurred in the past very frequently under certain conditions
will under the same conditions occur very frequently in the
future.

Thus induction (and probability behavior) is certainly justified if
the world is uniform. But we have said that X need not have a
guarantee of success for his individual action to be rationally justified.
By the same token, Carnap argues, we need not be *certain* of the
uniformity of the world for induction (and probability-behavior) to
be rationally justified. It is sufficient to establish that there is some
probability that nature is uniform. Many theorists argue that even
this much cannot be established, since it is an empirical statement
and could only be justified if we *presupposed* the method of induction.
But, for Carnap, all statements of AP probability are analytic,
including this one. It is in fact possible to establish by logic alone
that the hypothesis of the uniformity of nature is very probable (on
the accumulated total evidence) and that X therefore has a probability of success in the long run. Thus X is justified in his probability
behavior.

Now this appears to be a justification of induction by showing
something about the nature of the world: that it is probably uniform.
But there are a couple of catches. Catch number one is that if we
accept probable uniformity because Carnap has established it
analytically, then all we have accepted is a logical relation between
a set of evidence and a conclusion. But such an analytic a priori
statement can say nothing about the world – how then can it give
X an assurance of success in the real world? Carnap's reply is that
it cannot: 'it is not possible to give X an assurance of success even
in the long run, but only of the probability of success, as in [the
statement above]; and this statement is itself analytic.'[121]

Why should X act on the basis of a purely analytic statement?
Well, says Carnap, for the same reason he acts on an ordinary
probability statement, like the one that it is probably going to rain
tomorrow. In the first place, X has more than *just* the analytic
statement – he has also the empirical evidence mentioned in it. In
the second place,[122]

123

it is reasonable for him to take suitable action; for example, to take his umbrella or to bet on rain rather than non-rain. *For a practical decision is reasonable if it is made according to the probabilities with respect to the available evidence,* even if it turns out to be not successful. Going back to the general problem, it is reasonable for X to take the general decision of determining all his specific decisions with the help of the inductive method, because the uniformity of the world is probable and therefore his success in the long run is probable on the basis of his evidence, even though he may find at the end of his life that he actually was not successful and that his competitor who made his decisions in accordance not with probabilities but with arbitrary whims was actually successful.

So far as I can see, this amounts to saying that X is rationally justified in his behavior because it is reasonable to act on probabilities, including the Big Probability that Nature is Uniform. But if we knew why it was reasonable to act on probabilities we wouldn't have this problem in the first place. Thus, the first catch is that if we accept the probable uniformity of nature as analytic, we have no reason for acting on that probability except Carnap's bald assertion that it is reasonable to do so.

So much for catch number one – now for Catch-22 (Catch-22 is, of course, inescapable and all-embracing: 'There's just one catch, and that's Catch-22.'[123]) We have been arguing above about whether or not Carnap succeeds in justifying probability-behavior by merely establishing that the probable uniformity of nature is analytic. But whether or not this is a reasonable procedure, must we agree that he has established that nature is probably uniform? How does one go about this?

Well, first, we must codify all available empirical evidence and the hypothesis of high uniformity, then we compute the c^*-confirmation of the hypothesis given the evidence. Carnap asserts that this value will be high enough, though it would be reasonable to balk at that assertion (if the hypothesis were stated as a universal law, for example, its confirmation would necessarily tend towards 0 as a limit[124]). But even if we accept that the value of c^* is high, we need not necessarily accept that the probability of h is high (*vide* the previous section). Thus Carnap has at most shown that probability-behavior is rational *if we first accept his theory of probability.* Catch-22 is that this justification, like so many others, requires us first to assume what we are setting out to prove. It is, in short, viciously circular.

So much for Carnap's views in *LFP*. I have quoted them rather extensively, in the process of criticizing them, because he says later that he was not attempting a justification at all! Instead, the section from which we have quoted is said to have been 'mistaken' for such a justification because of his 'misleading' use of the word 'justification.'[125] In 'Replies and expositions,' Carnap agrees with Burks's assertion[126] that the controversial section had dealt only with the 'internal' question of the c^*-confirmation of the uniformity hypothesis, while the justification of induction (or the selection of a c-function) is an 'external' question which cannot properly be answered by such a method. On this latter question, Carnap claims, he 'said next to nothing' in *LFP*.[127] Even in 'Replies and expositions' he refuses to take up the general problem, saying only that such a justification *will not* require general synthetic propositions about the nature of the world because 'questions of rationality are purely a priori'[128] in the sense that rational procedures are justified only by showing that they conform to our intuitive notion of rationality – not by showing they conform to the nature of the world (see above, 'The metaphysical status of *P*'). In a sense there can be no *question* about the rationality of probability-behavior in general, just as there can be no question about the rationality of rational behavior in general. The only question is whether certain theories or types of actions conform to our a priori standards of rationality. Carnap thinks that actions based on the c^* (or similar) system of quantitative inductive logic do conform to those standards, but he chooses not to argue this systematically. He does, however, present a rather extensive argument intended to show that AP theories of probability are far superior to RF theories as a 'guide to life,' which can be taken to mean that action based on AP probability is at least *more rational* than that based on RF probability.

In his argument, Carnap says that a statement of probability$_2$ (= RF probability), while having the advantage of saying something about the world and hence being empirically meaningful, suffers the disadvantage of all empirical statements in being sometimes false, often unknown, and never certain. Probability$_1$ (= AP probability), on the other hand, is analytic and therefore knowable with certainty, but says nothing about the world. Yet Carnap maintains that it can and should act as a guide for action. In an analogy with estimates of other kinds, he says[129]

> Practical decisions of a man are often dependent upon values of certain magnitudes for the things in his environment. If he does not know the exact value, he has to base his decision on an estimate. This estimate is given in a statement of the form: 'The estimate for the magnitude in question with respect to such and such observational results is so and so.' This statement is purely analytic. Nevertheless it may serve as a basis for the decision. It cannot, of course, do so by itself, since it has no factual content; but it may do so in combination with the observational results to which it refers.

Now clearly Carnap is correct in saying that from the evidence together with the estimate statement we cannot infer the *actual* value of the magnitude. Why then do we act *as if* we knew that value? We seem to act in accordance with some rule of the sort that I call the Rule of Estimation:

> RE: When a magnitude is unknown, assume it to be (posit) that value indicated by the rules of estimation and all available evidence.

Those who do not care for rules of action might prefer merely to give arguments justifying 'estimation behavior' as rational. Such an argument must say something about the nature of the world and something about the rules of estimation. It should be clear that what makes an estimate more or less useful is not the depth of its analyticity, but the extent of agreement between the rules of its derivation and the external world. If we chose to estimate the true value of a quantity, X, by the sum of the squares of the observed instances, X_i, rather than by their arithmetic mean, we would still have an analytic estimate, but one that would be thoroughly useless. Thus Carnap's own analogy is strongly suggestive that any concept of probability – analytic or not – stands in need of a justification by some uniformity principle or by a Reichenbachian or other pragmatic argument.

Although Carnap sees there is a problem here, he thinks of it as a problem in the *application* of probability only. He in fact discusses many possible rules, similar in form to my RE, for the application of probability$_1$ to practice. He concludes that the optimum rule for action is[130]

Rule R_5. Among the possible actions choose that for which the estimate of the resulting utility is a maximum,

where the value of an action is its mathematical expectation and its utility is calculable according to Bernoulli's Law of Utility.

It is Carnap's position that 'It may be assumed that the present rule or a similar one using likewise inductive concepts like probability$_1$, estimate, etc., would be adequate as a "guide of life", that is, as an explicatum for the vague concept of a reasonable decision'[131] provided that adequate methods for the quantitative determination of probabilities and utilities were available.

Now this rule (and most other rules for the same purpose) depends fundamentally upon the notion of a mathematical expectation. But Carnap has a rather compelling argument to the effect that a definition of 'mathematical expectation' which is based on probability$_2$ is considerably different from the historical meaning of that term. As originally developed, the expectation function was intended to serve as a guide to action by projecting an expected value for each alternative course *on the basis of present evidence.* What Carnap calls the c-mean estimate of a value will still serve that purpose. But a 'mathematical expectation' based on concepts of probability$_2$ (RF probability) is somewhat different. Such a value is characterized by a dependence not on the *evidence* but on the *facts.* In any empirical situation, it is quite likely that a person will not know what the correct value is: thus it hardly seems appropriate to call it his 'expectation' or to urge it as a guide for his action. Perhaps it would be acceptable to call it his 'objective expectation,' since it is based on the actual values and frequencies of the situation (this would be particularly appropriate if we agree that there is such a thing as *the* probability of X), but to avoid unnecessary proliferation of terms, we shall adopt Carnap's convention and refer to the two concepts as 'expectation$_1$' (Carnap's c-mean estimate, or, generally, a priori expectation) and 'expectation$_2$' (any relative frequency expectation).[132] It is clear that an expectation$_2$ cannot be used as a guide to life, since it is generally not known at the time of decision. Expectation$_1$ must be used instead. Thus probability$_1$ (AP probability) is the only appropriate guide for life, and behavior based on it is more rational than that based on probability$_2$ (RF probability).

11 CHIEF CRITICISMS OF A PRIORI THEORIES

The strongest single criticism against AP theories is that the number of logically possible outcomes in no way determines the probability of one of these outcomes in the real world – no merely logical division can tell us what will probably happen in the future.

This is, of course, the same objection that was raised most strongly against the classical conception of probability. The reader may be understandably surprised to find it recurring here, since Keynes and Carnap both disavow the unrestricted application of the Principle of Indifference which characterized the Classical school. But remember that despite his restrictions and revisions, Keynes ended up retaining the Principle of Indifference as the only legitimate source of numerical initial probabilities. Thus, even if he had succeeded in resolving *every* internal paradox, contradiction, and absurdity which can be squeezed out of the Principle, he could still be criticized on the grounds that it is a *fundamentally* invalid method.

It might require a closer look to see that Carnap is likewise subject to this attack, since he explicitly rejects the Principle of Indifference on the grounds that it 'leads sometimes to quite absurd results and in its strongest form even to contradiction....'[133] But Carnap's preferred function for degree of confirmation, c^*, in the end declares that every structure-description is equally likely and within the structure every state-description is equally likely. Carnap denies that this assignment is based on or justified by the Principle of Indifference, but the *effect* is the same – probabilities are determined by counting (two stages of) equipossible cases. If the objection is valid in its particular form, that counting possible cases (equally) gives us no valid measure of probability, then it tells against Carnap as much as against Keynes and the Classical theorists. And of course, if it is valid in its general form, that no merely logical division can determine probabilities, it discredits *every possible* AP theory.

The basic response to this criticism is that AP probability deals with propositions and properties, not things. It doesn't prescribe to the world because it doesn't say anything about the world *But* of the many possible a priori probability systems, some will resemble rather closely the relationship between natural events – just as some systems of deductive logic and some parts of mathematics resemble the structure of the world. When this happens, we have a tool to use in experience – a partial model of the world. If we then wish to know the consequences of a real event, we ask the system for the modelled consequence in deductive logic, mathematics, and probability theory. We then correlate the modelled consequences with another real event, and act on that expectation. Evolutionary and pragmatic adjustments in these systems have brought two of them into very close agreement with the world – it is now our task to systematize and refine our probability intuitions to bring them also into agreement with reality.

Besides this general response, there is also a rather forceful *ad hominem* response. Most of these criticisms of AP theories are advanced by people who claim to be 'scientific' and 'empirical' in their outlooks – RF theorists and working scientists and statisticians especially – and who therefore claim to oppose anything a priori as anti-empirical. Yet these very same people will heartily praise mathematical modelling as a tool – perhaps the most important tool – of science! So far as I can see there are three chief reasons why these people reject the attempt to build a mathematical model of induction: (1) they think they already have an adequate enough model in the RF theory of probability, (2) they find it very difficult to accept such 'counter-intuitive' results as 'The initial (null) probability, that a man is a leper is 1/2,' because (3) they persist in taking such probability statements as saying something about the world.

The adequacy of RF theories will be examined later. On the 'counter-intuitive' statement, the first necessity is for a sympathetic understanding of what it *means*. Properly rephrased, it says only that the logical width of the predicate 'leper' is 1/2 that of the predicate 'man.' It does not mean that we should reasonably expect half of the men we meet to be lepers. If this is not sufficient to remove the difficulty, we should remember Carnap's warning that 'isolated intuitive judgments are very often unreliable.... [intuitive] judgments are more useful if they are made, not on isolated points, but in the context of the tentative construction of a system.'[134] Those of us who remember our undergraduate doubts about material implication should be sympathetic to this rejoinder. Intuitions are immensely valuable, but they can be misleading. No pragmatically successful algorithm should be rejected just because parts of it are counter-intuitive.

The point of these arguments is that empiricists should beware of a knee-jerk rejection of everything 'a priori' based on unhappy experiences with rationalistic metaphysics. The abstract formal systems of logic and mathematics have served us faithfully – the abstract formal systems of AP probability should be given a chance to do so as well.

So much for the major objection to a priorism as such.

The second general objection to AP theories is that they are odiously subjective because they recognize probabilities relative to given evidence only. Many RF theorists consider this to be a grave defect in AP theories, rendering a probability subordinate to the state of knowledge. Their theories, they maintain, deal with the *real* probability, which is objectively determined and not relative to anything.

Both Keynes and Carnap explicitly denied the charge of subjectivity. To say that probabilities are subjective is to say that they depend on people's thoughts – this is what both deny. But to say that probability is a logical relation between evidence and conclusion is to say that it has *something* to do with people's thoughts. For AP theorists, probability theory has a *regulative* relation to people's thoughts – it doesn't describe the way they actually do think, it describes how they must think if they are to think correctly. Since this still has to do with thinking, Carnap is willing to call it a qualified psychologism,[135] but both he and Keynes stress that it is nevertheless objective in precisely the way that rules of logic and mathematics are. '$P(A \vee B) = P(A) + P(B) - P(A\&B)$' is just as true as '$2 + 3 = 3 + 2$' and whether or not Bob Schultz believes it has no effect at all. A thing is not probable, says Keynes, because we think it so.

But even if we grant that AP theories are not 'subjective' in the simple, straightforward way that Subjectivist Theories of Probability are, doesn't the evidence-relative feature make them more subjective than the RF theories, which recognize one and only one empirically determined value independent of the state of our evidence? In one sense this is obviously and significantly true. The *aim* of RF theories is a complete description of certain repetitive aspects of empirical reality. The *aim* of AP theories is a complete description of the (actual? optimal?) method used by human beings for probability inferences. Clearly the latter is more anthropocentric than the former, and, to that extent, more 'subjective.'

Again, each simple statement of RF probability purports to assert a fact about the empirical world, while no AP statement does so. This makes it more objective if (but only if) we would also say, for example, that a statement of descriptive biology is more objective than a statement of theoretical mathematics.

Finally, it might be supposed that AP theories are more subjective because they make the probability of an event, *E*, dependent upon the state of our knowledge, while RF theories do not. This is false. For any event, *E*, there are many AP probabilities of its occurrence, each relative to a given evidence statement but *none relative to human knowledge*. Our problem in dealing with *E* is to select the most appropriate of these probability statements to use as a basis for action. This selection does indeed depend upon the state of our knowledge and the keenness of our judgment, but the probability itself does not.

The situation with RF theories is almost precisely parallel. There are many real RF probabilities of *E*, depending on the choice of a reference class.[136] This choice is likewise dependent on the state of our knowledge and the keenness of our judgment. So far there is no difference between the two. Now suppose we have selected our reference class (parallel to choosing the evidence-set in AP theories). We have indeed picked out a single value for the probability of *E*, *but we don't know what that value is*[137] (whereas we *do* know the AP probability determined by *e*). The only way we could possibly know that value is if we had a complete description of the reference class, or at least a precise value for the relative frequency of occurrence of things like *E* in the reference class. In the first case, the probability is known to be either 1 or 0 by either theory. In the second case, where the frequency is known but not the particulars, Carnap's Direct Inductive Inference (and Keynes's methods, in some cases) would give precisely the same value to the probability. In either theory, the value of $P(E)$ is not *determined by* the state of our knowledge (the AP probability is determined by a logical relation; the RF probability is determined by an empirical frequency) but it is only accessible to the RF theorist at a given level of knowledge and is equally accessible to the AP theorist at the same level of knowledge. So far no difference.

Now we come to the more usual case. Suppose we have identified our reference class (chosen our evidence set) and thereby singled out 'the' probability of *X*. This probability is known to the AP theorist, since it involves only logico-mathematical calculations. But it is usually *not* known to the RF theorist! He can't know what the relative frequency *is* unless the series has exhausted itself. For most series this is not the case and for infinite series it is never the case. But presumably he must assign some value to $P(E)$, in the interest of action. The value he selects will be totally dependent on our experience of the initial segment of the series – how else could he arrive at a value? Therefore the RF probabilities used in practice are (almost) never the true values determined by Nature, but approximations only, which are entirely dependent upon the state of our knowledge. Whether this counts as a strong objection to RF theories will be discussed later[138] – we brought it up here only to demonstrate that the sense in which the AP value of $P(E)$ is dependent on human knowledge is not a peculiar subjectivist defect in that theory, but is the same sense in which most of our empirical knowledge of the world, including especially the RF value for $P(E)$, is dependent on past human experience.

Another general criticism of AP theories is that they are never falsified by experience and hence can never learn from experience.

We mentioned earlier the presumption originating with CTP and generally carried over in Keynes that methods based on the Principle of Indifference work when and only when conditions are standard (the dice are fair, etc.).

A broader general explanation is that AP probabilities are never refuted by experience because they are indissolubly linked to the evidence on which they are based.

Suppose someone hands us a die and asks us the probability of rolling a Five. Using AP theories (or almost any other) we would naturally reply that $P(5) = 1/6$. Now we roll the die 100 times and invariably get a Three. Haven't we been proven wrong?

The RF theorist who had (perhaps inappropriately) agreed that $P(5) = 1/6$ must indeed now admit that he had been mistaken before, but now (he claims), he knows that $P(5) = 0$. But a careful a priorist would originally have said '$P(5, e) = 1/6$.' Now he will say '$P(5, e') = 0$.' The first statement asserts that the probability of a Five is 1/6 on the originally available evidence, e; the second, that the probability is 0 on the evidence, e', available after the series of rolls. Both statements are true and can be uttered *at any time* without being mistaken. As it happens, we are interested in and rely on one before the experiment and the other after, but that does not mean experience has falsified either one. It means that experience has made interesting and important an equation that had previously been uninteresting and unimportant, as happens continually in applied mathematics.

At most we can say that experience has taught us that our earlier knowledge was misleading as a base for prediction – something which happens with regrettable frequency in all human endeavors. It remains true that the probability based on what we knew then was 1/6 and that fact cannot be altered.

These three criticisms – that logic is useless for prediction, that AP theories are subjective, and that they cannot learn from experience – are the main general objections raised against AP theories. Now let us look briefly at some specific criticisms of Keynes and Carnap.

If the number one criticism against Keynes is that he *does* employ the Principle of Indifference, the number two criticism might be that he *only* employs it – he frequently says that numerical probabilities can be obtained only where the Principle is applicable. This means that such cases as biased dice, and, especially, actuarial cases with regular frequencies, cannot be adequately covered by Keynes's theory. This is a point that we saw told heavily against the Classical theory. Keynes tried to solve this problem by developing his methods of statistical induction. These methods have been largely ignored, and, I think, deservedly so. The essential spirit of Keynes is that all probabilities are non-numerical, and many non-comparative, except in those cases where it is legitimate to apply the Principle of Indifference.

Besides this shortcoming, Keynes is also dated epistemologically. Few philosophers of the present Anglo-American generation have much sympathy for a system in which axioms are part of the a priori laws of thought and even particular probability judgments are based on some form of direct perception or intuition.

These criticisms have been partially responsible for the decline in acceptance of Keynes's views, but in a very real sense I think that Keynes has not so much waned as been eclipsed. Those philosophers who are inclined towards a priori theories now find a much greater center of attraction in the astonishingly creative work of Rudolf Carnap. The fact that Carnap's system is flawed is, as Kemeny says,[139] or less importance than the fact that a system of quantitative inductive logic has actually been constructed. The urge to tinker with the system (or try to outdo it with a better one) has proven to be strong and widespread among logicians. Even Hilary Putnam, who has his doubts about the entire enterprise and thinks it is demonstrably impossible to construct a 'best possible' algorithm for induction[140] has taken the time to contribute an improvement to Carnap's system.[141]

After the general objections to any a priori theory, the most frequently heard complaint about Carnap's theory is that it makes probability language-dependent, and, at least in the present form, dependent on very inadequate languages at that.

To forestall a possible confusion, let me remark that some measure-functions are affected by language-changes outside the range of the immediate evidence and hypothesis while some are not. The first group might well be thought to be language-dependent in a more invidious sense than the second. But in both cases the value of the *c*-function is determined solely by the structure of the language: its number of structure-descriptions, the range of its predicates, etc. It is this more general language-dependence that is the point of the most interesting objections, not the sensitivity of some measure-functions to external changes.

To continue then with the main point. Carnap's system bases probabilities in the language rather than in the world – is this reasonable or mistaken?

In *LFP*, Carnap sought to mitigate this problem by requiring that the language of inductive logic be completely adequate for describing the world, thus, in effect, reducing the gap between fact and language. He states this[142] as the

> *requirement of completeness*: the set of the [predicates] in a system *L* must be sufficient for expressing every qualitative attribute of the individuals in the universe of *L*, that is, every respect in which two positions in this universe may be found by observation to differ qualitatively.

This requirement has the effect of *fixing* the number of predicates in *L*. As a result, although *c*-values are still *determined* by the language, they are not subject to any *variation* in the language, because variation is no longer allowed. Furthermore, the number of predicates is determined by the nature of the universe – probabilities are indeed connected to the real world, though only indirectly.

But even if this requirement abates somewhat the language-dependence of *c*, the cure, as Ernest Nagel suggests,[143] may be no improvement on the disease:

> Unless we do have good reasons for fixing the number of primitive predicates in a complete set, we cannot, even in

134

principle, calculate the value of c^* for non-trivial cases, so that the inductive logic based on c^* is simply inapplicable. But the assumption that a complete set of primitives contains a given number π of predicates is not a truth of logic; it is at best a logically contingent hypothesis which can be accepted only on the basis of empirical evidence. The assumption is not a logical truth, for it in effect asserts that the universe exhibits exactly π elementary and irreducible qualitative traits, into which all other traits found in nature are analyzable without remainder. It is an assumption which would be contradicted by the discovery of some hitherto unnoted property of things (e.g. an odor or distinct type of physical force) that is not explicitly analyzable in terms of the assumed set of basic traits.... it is difficult to avoid the conclusion that the assumption that we have, or some day shall have, a complete set of primitive predicates is thoroughly unrealistic and that in consequence an inductive logic based on that assumption is a form of science fiction.

Criticisms of this sort, plus his own intuitions, led Carnap to be unsatisfied with the requirement of completeness, but he could not dispense with it until later, with Kemeny, he developed an inductive logic of somewhat different type, in which the number of predicates need not be fixed. This later system is beyond our present scope; but even if it eliminates the requirement of completeness and the dire consequences Nagel attributes to it ('science fiction,' indeed!), it still remains true that probability values are established by features of the language rather than features of the empirical world. (We make throughout the simplifying but false assumption that language is not part of the empirical world.) Can this be tolerated? Can we hope to say anything about future experience by merely counting linguistic divisions?

At this point we almost seem to have reduced the 'language-dependent' objection to the 'anti-a-priori' objection above, because if we consider our language to be part of our logic, objecting to grounding probabilities in language amounts to objecting to grounding them in logic (if we instead abandon our simplifying assumption and consider language to be part of the world rather than part of logic, it follows that a language-dependent probability theory *is* based on the real world). Then the parallel solution is: language-dependent theories are useful if and only if the language in question closely resembles the real world[144] (and the non-linguistic or additional features of the system such as the axioms are adequate, of course).

This leads naturally to the next objection: even if language-dependence is not bad *per se*, the languages for which Carnap (and others) have defined *c*-functions are childishly simple, artificial in nature, and woefully inadequate for the development of modern science.

This objection must be admitted, so far as it goes. But look how much can be accomplished with even these 'childishly simple' languages! All of the results of the Classical theory (and Keynes's quantitative theory) are immediately available. Any statistical frequency of simple property coincidence adequate for RF theory will give an (almost) identical value in Carnap's theory. So the theory successfully accomplishes most of what its rivals can do (whether it can do everything the RF theory can do is still a matter of technical debate). In addition, it promises to formalize the judgments of probability and induction which we make every day on the basis of evidence which is unsatisfactory for other types of theories. That it is still a long way from accomplishing all of its goals is a fact. But, as Putnam says,[145] 'who in 1850 could have anticipated the evolution of deductive logic in the next hundred years?' Perhaps the child-like simplicity of the languages *L* should not be overly stressed in criticizing Carnap's system since it is, after all, a mere child in length of historical development. And, to paraphrase Dr Johnson, 'The amazing thing is not how good or ill a quantitative inductive logic he constructs, the amazing thing is that he constructs one at all.'

Another objection which Carnap says has led many 'empiricists' (presumably logical positivists) to reject AP probability theory is that the theory seems to give rise to unverifiable sentences. An example is 'On the available meteorological evidence, the probability that it will rain tomorrow is 1/5.' We can verify that it *will* rain tomorrow (or not) by observation tomorrow, but there is in principle nothing we could observe which would verify that the probability of rain is 1/5. Therefore the statement violates the Verification Theory of Meaning and is meaningless.

Carnap agrees that this sentence cannot be verified by the impossible observation of 1/5 rain tomorrow, or by any other empirical observation. But, he says, the same is true for the sentence

'If there will be rain and wind tomorrow then there will be rain tomorrow.' Here, again, no observation of tomorrow's (hypothetical) rain is necessary or even relevant for verification of what is asserted – because what is asserted is not a fact about rain tomorrow but a fact about the logical relations between sentences. Statements of inductive logic, like those of deductive, are verified by logical rather than empirical investigation. If this characteristic is not sufficiently deplorable to reject deductive logic, why should it cause us to give up inductive logic? The objector apparently fails to understand either *what* is being asserted in such singular predictions or *the grounds* of such an assertion.[146]

Still another objection to Carnap's method is that we can't possibly quantify all of the factors which enter into inductive judgments – especially such amorphous factors as analogy and variety. Ernest Nagel especially presses this argument – see *Principles of the Theory of Probability*, pp. 68 ff., for example – and Carnap admits its seriousness.

Actually, Carnap does have a method – the method of Inference by Analogy – for measuring the confirmatory effect of positive analogies. It is the principle of Negative Analogy, or Variety of Instances, which remains to be systematically represented. The basic idea of negative analogy is perhaps best given by example. If our hypothesis is that all animals which have hearts also have kidneys, we will normally think it better confirmed by successful examination of a dog and a gorilla than by examination of two dogs. Each time we add a new species to our list, that varied instance counts for more than a repetitive one. What we lack is a means for saying *how much more* varied evidence should count. Nagel thinks such a measure cannot be satisfactorily constructed. Carnap thinks it can but we just haven't had time to do it. Considering the relative success Carnap had with positive analogy, I see no reason to doubt that his successors could eventually develop some intuitively appealing measure of negative analogy. To do this also for such factors as simplicity, agreement with established theories, fertility in new predictions, etc., is indeed a soberingly difficult problem, but it does not seem to be a priori impossible to develop these measures. Whether it will be *practically* impossible, or whether they would in turn make inductive calculations so complex as to be practically impossible are questions which cannot yet be answered with authority – each must judge according to his own lights.

A. J. Ayer has repeatedly objected to AP theories on the grounds that what we really want is $P(h)$, not $P(h, e)$. I discussed this objection and replied at some length in the first part of the section 'Absolute probability and physical chance,' and will not rehash the matter here. But in the course of his argument against relative probabilities, Ayer raises another objection against logical probabilities in general:[147] if all are equally valid, how do we choose one to act on?

> For, if we are presented only with a stock of necessary facts to
> the effect that certain statements, or groups of statements,
> bear logical relations to each other in virtue solely of their mean-
> ing, I do not see what reason there could be for differentiating
> between the items of this stock as bases for action.

If we leave aside the question of the rationality of acting on a priori truths in general, and focus instead on the question of *which* a priori truth to act upon, as Ayer seems to intend in this passage, the resulting problem is at best a methodological difficulty – certainly it is no insuperable bar to action. At first one may be dazzled by Ayer's image of an infinite array of a priori truths, each valid for all time and each a candidate for the honor of guiding our action ('Choose me!' 'No, choose me!'). But a little reflection reveals that the situation is no different from that in mathematics or deductive logic. If we wanted to know how many pieces of fruit are in a sack containing three apples and two oranges, we *might* stand transfixed in awe at the infinitude of valid mathematical principles at our command. And if we wished to know whether all red apples are apples we *might* profess our inability to choose from the vast 'stock of necessary truths' which constitutes deductive logic. But anyone who actually hesitates in such cases must be lost indeed. The fact is that it is *usually* quite obvious which a priori principle is relevant in a particular case. That it is not *always* quite obvious is a fortunate circumstance which guarantees the continued employment of logicians and mathematicians; but it certainly is no reason for abandoning logic or mathematics.

Similarly, in the selection of an AP probability statement, we sometimes use something as simple as the assumption that a certain die is fair as our evidence. We sometimes consider everything known about a certain situation, and we often focus on items which we take to be especially relevant. It is no doubt true that there can be no one algorithm for arriving at well-founded judgments of applied probability, but this defect applies equally to any theory of probability (or mathematics or logic or science) and is no special criticism of AP interpretations.

If one wishes to be more specific in one's criticism, there are many possible objections which can be raised against c^* as a particular c-function, such as the much-lamented fact that a universal statement in an infinite universe always receives zero confirmation: all predicates must be independent of each other; we must have names for all individuals; values of c^* may be unreasonable if there are many predicates, and so on. Some of these problems have been overcome by the work of Kemeny, Putnam, Hintikka, Bar-Hillel, *et al.*, and by such later works of Carnap's as *The Continuum of Inductive Methods*, 'Replies and expositions,' and *Studies in Inductive Logic and Probability*. Readers who are interested in these technical details should study the works in the bibliography.[148] Whether or not c^* suffers from irremediable technical difficulties is not the kind of problem which concerns us here. Instead, let us consider the much more intriguing question of whether there *is* a best possible c-function – whether we can *in principle* ever hope to develop a confirmation-function which does the best possible job of generating probabilities. Hilary Putnam has argued that we cannot.

Putnam's argument[149] is of the general type which logicians call a 'diagonal argument.' Put very simply, it asserts that if we have a c-function, c, developed for a language adequate for science (including, especially, sequential enumeration of individuals or events) then it is always possible to construct a world in which c fails. ('If N is the first point at which Red is predicted, let N be black. If $N + M$ is the next point... let $N + M$ be black.') This sounds a bit like cheating to some non-logicians, since we, in effect, ask what c is going to predict before we decide what the real world is going to be like, then decide in a way to contradict c. It is a rather *malicious* selection of worlds, but it is a selection of a *possible* world, and the fact that such a world is possible shows that c will not invariably lead to success. Furthermore it is possible to construct another function, c', which duplicates all the successes of c *and* succeeds in predicting redness ('N is red if c predicts N is not-red'). Thus c' is 'better' than c in this world. But the diagonal method can then be applied to c' to show that there must be a better method, c'' which succeeds in at least one world where c' would fail. Since the method is quite general, we arrive at the conclusion that *there can be no one best c-function for all possible worlds.* Since we don't know *which* of the possible worlds we live in, it follows immediately that it is logically impossible to establish one best c-function *even for our own world.*

In his published reply to Putnam,[150] Carnap rejects the conclusion and attacks Putnam's definition of a criterion of 'success', but I think he fails completely to respond to the intuitive notion. In the preface to *LFP* he admits[151] that for every c-function, c, there is always *some* world in which another function is better. There is even, for *any* given function, at least one world in which that function is better than c. But Carnap thinks[152] we can reasonably judge between functions on practical grounds:

> Suppose, for example, that in comparing two given inductive methods we find that the number of those state-descriptions in which the second method is more successful is a million times as large as the number of those in which the first method is more successful. Then it may well be that this result would influence us against regarding the first method as more adequate than the second and against choosing the first in preference to the second for determining our practical decisions in the actual world, whose total structure is not known to us and for which we therefore cannot know which of the two inductive methods would be more successful in the long run.

But suppose Carnap's first method, c_1, is more successful N times, while c_2 is more successful $1,000,000$ N times. Putnam's argument is that he can always construct c_3 which is more successful in at least $1,000,000$ $N + 1$ worlds. (Make c_3 identical to c_2 for the first $1,000,000$ N worlds, construct world $1,000,000$ $N + 1$ in which c_2's predictions in some series are always false, and c_3 detects that fact and predicts accordingly.)

I happen to agree with Putnam's claim. In fact, I used a very similar (though non-general) argument against Reichenbach's justification of induction in my dissertation.[153] But the fact that such worlds can always be logically constructed gives me only the faintest of fears that they are real. Thus, they have almost no effect on my attitude on induction. There is no need to fear that we will make infinitely many *important* mistakes by embracing the wrong c-function, because nobody (or certainly not everyone) is so devoted to a method that he always sticks with it despite a uniform lack of success.

If we put the objection another way, Putnam has shown that it is always logically possible to improve upon a *c*-function, therefore we can always be sure that we don't have the best. But not having the best is seldom a good reason for not having any. My car is imperfect, but I don't eschew automobiles. If someone develops a system of quantitative inductive logic which simply and adequately reproduces our intuitions except where it demonstrably improves upon them, I will not restrain myself until a 'perfect' system comes along, and neither will Putnam if I know him at all well.

One final objection which Arthur W. Burks has raised may be appealing to linguistic or ordinary language philosophers. It is just that 'there is an important difference in meaning between "probability" and "*c**" in belief contexts.'[154] This difference, to which we have already alluded, is that one who is acquainted with Carnap's work and is logically competent might be able to compute *c**-values, while disputing that they have anything to do with probability. He might, for example, agree that $c^*(h, e) = r$, while denying that $P(h) = r$ [or even $P(h, e) = r$].

This difference in meaning, Burks suggests, might be accounted for by the fact that *c** is intended as an explication, rather than an analytic definition, of 'probability' and that it can therefore quite properly differ in meaning.

A final reply, of the 'So's your old lady' or *tu quoque* type, would be that *none* of the major theories of probability offers definitions that are equivalent in meaning to the ordinary word 'probability', and thus AP probability is not especially discredited by its failure to do so.

12   CHIEF VIRTUES OF A PRIORI THEORIES

Richard von Mises spoke contemptuously of Classical theories of probability because they apply to at most a very small fraction of real situations. All biased dice and actuarial situations seem forever beyond their ken. But even adding von Mises' long empirical series to the Classical fair games of chance doesn't exhaust the universe, or even the everyday application of 'probability'. It is the great virtue of the AP interpretation that it claims to cover *every* case in which we do (or might) think of the probability of something (though Keynes's version will generally not give us a *number* for the probability). This virtue is partially shared only by the Subjectivist Theory, among the major contenders. It gives credence to Carnap's contention that *AP theories come closest to explicating 'probability' in the fullness of its meaning.*

Carnap claims that the word 'probability' has historically referred to the explicandum probability$_1$ and this explicandum is normally explicated by an a priori theory of probability.[155] Probability$_2$ 'goes back not more than about a hundred years.'[156] It originated when the *estimate* of a frequency – a valid probability$_1$ concept – became confused with the *frequency itself* and the word 'probability' was incorrectly applied to the latter. This gives AP theorists an historical basis for their claim that RF theories are not theories of *probability*. Nevertheless, much of the serious writing on probability during this century has been done by mathematicians, statisticians, and scientists who are using the word 'probability' to refer to relative frequency in a population. This body of usage is so considerable and so intellectually respectable that it is no longer plausible to argue that such people are simply misusing the word 'probability.' Thus we cannot ignore probability$_2$ in a discussion of probability, but we can recognize that probability$_1$ takes *historical* and *etymological precedence* over it.

I have argued above that the chief distinction between theories of probability is in the definition of 'probability' and the resultant method for obtaining *initial* probabilities (once these are available, their manipulation by means of the probability *calculus* varies little from theory to theory). The AP theory, then, asserts that these initial probabilities exist in far greater profusion than its rivals allow – but this means that the calculus of probability will be much more widely applicable than its rivals allow. Therefore we can say that AP theories *expand the domain of the probability calculus.*

Furthermore, this definition and method lead to the result that every probability, if it can be known at all, can be known *now*. We have only to consult our intuition or perform some calculations to determine what any probability is. This means that we can know the probability of an event before it (or a series of similar events) has occurred. AP probability, therefore, is a much more comprehensive *guide to life.*

But not only can each probability value be known in advance, it can also be known *precisely*. That is, if we agree on some effective AP method, a probability value determined by that method is the true value, by definition. An RF theorist, on the other hand, can never know if his measurements have given him the true value of the frequency, nor can he establish any algorithm which will guarantee to do so. AP theories have the advantage, then, that, if a theory is accepted, *the measurement problem is solved.*

The final major advantage of AP probability is that it *makes inductive logic possible.* By specifying a confirmation-measure between sentences which can vary from contradiction to implication, AP theories allow us to talk about evidentiary relations which are less than conclusive. This is something no other theory can do.[157]

It is true that at present these virtues are more potential than actual. It cannot be said that many practicing statisticians or handicappers compute probabilities by counting predicates in a language. Whether anyone ever will do so remains to be seen. But at least the first steps have been taken. As Kemeny says:[158]

Carnap has taken this fundamental problem, of the method by which inductions are reasonably performed, out of the stage of fruitless debate, and he has shown us the way to constructive research. He himself achieved the first important result in this research. Therefore, we must class Carnap's contribution to the problem of induction among the greatest achievements of modern Philosophy.

IV

RELATIVE FREQUENCY THEORIES OF PROBABILITY

The major leading opponent of the A Priori theory of probability is the Relative Frequency (RF) theory.

While Nagel is willing to trace its roots back to antiquity ('this view already appears in Aristotle, was proposed by Bolzano and Cournot during the last century and further developed by Ellis, Venn, and Peirce, and was finally made the basis for a subtle mathematical treatment of the subject by von Mises and other contemporary writers'),[1] the safer course is to credit John Venn with the first serious proposal to identify probability with a relative frequency (1886) and Richard von Mises with the first systematic development of that theoretical position (1928).[2] Hans Reichenbach is perhaps the most philosophical of the frequentists and may fairly be taken as the prototypical RF theorist.

1 BASIC IDEAS

The most important basic idea of the RF theory is the metaphysical asertion that (1) a probability value is a measure of an empirical, objective, physical fact about the external world. This is directly contrary to the doctrine of the a priori theories (and, to some extent, the Classical theory) which contend that probability is a logical feature of certain situations or their descriptions. It is even more violently opposed to the subjectivistic thesis that probability is an index of human beliefs or attitudes. It is intended, indeed, to make

144

probability theory either a part of descriptive physical science (von Mises) or, at least, a part of the theoretical structure of physical science (Reichenbach).

From this fundamental conception of probability as an objective property of the world follows immediately RF probability's second basic principle: (2) A probability value is never relative to any evidence but is uniquely determined by the nature of the real world (the fact of the matter; the state of affairs which obtains, etc.).

Like the first principle, this idea stands in direct contrast to the a priorists, who assert that every probability statement must contain essential reference to the evidence and the subjectivists who contend that such statements must refer essentially to the human being who believes them. This is, of course, a necessary consequence or elaboration of the fundamental principle: since probability is an objective property (like mass, they say, or magnetism) it cannot be the case that human beliefs (whether factual *evidence* or subjective *attitudes*) can have any influence whatsoever on its value. So much is clear from an analysis of the basic concept.

The third RF principle is: (3) All probabilities can be known a posteriori only. Again the principle flows smoothly from the basic conception of probability as a property of the physical world. All empiricists agree that such properties are knowable a posteriori only. Even philosophers who accept the possibility of synthetic a priori knowledge generally think that particular physical properties are contingent and therefore only knowable a posteriori. For example, we *may* know a priori that an object will have a Euclidean shape and that it cannot be red and blue all over, but we cannot know a priori what shape it will have or which color it will be. It follows that if such philosophers accept the definition of probability as the relative frequency with which one contingent property is associated with another contingent (and independent) property, they must likewise agree that such probability values can only be known a posteriori (we ignore the limiting values of 1 and 0 for the sake of simplicity).

Many RF theorists claim that this epistemological principle places probability theory squarely on a par with physical science as the less than certain but useful and generally reliable ideal of knowledge about the external world. Critics have urged as epistemological consequences of RF theory that not only can such probabilities not be certainly known, they cannot even be known to exist, and

145

hypotheses concerning their values are neither confirmable nor disconfirmable.

2 DEFINITION OF PROBABILITY

At the beginning of his philosophical work, *Probability, Statistics, and Truth*, Richard von Mises follows the traditional method of citing several uses of 'probability' and quoting several definitions for that term. However, he soon makes it clear that he does *not* plan to accommodate such definitions. 'Let us now consider,' he says, 'a way by which we may arrive at a better definition of probability than that given in the dictionaries, which is so obviously unsatisfactory for our purpose.'[3]

The method von Mises chooses to employ he calls 'the scientific method of developing new concepts.'[4] It has two characteristic features:

in the first place, the content of a concept is not derived from the meaning popularly given to a word, and it is therefore independent of current usage. Instead, the concept is first established and its boundaries are purposely circumscribed, and a word, as a suitable kind of label, is affixed later. In the second place, the value of a concept is not gauged by its correspondence with some usual group of notions, but only by its usefulness for further scientific development, and so, indirectly, for everyday affairs.

This process is so nearly independent of ordinary language and results in concepts which are so new or different that 'The person who arrives at a new scientific concept may be inclined to invent a *new* name for it.'[5] Quite likely he will choose a related classical or foreign term for his concept to emphasize its independence from the colloquial term. One gathers that von Mises would like to do the same for 'probability,' and is stopped only by the paucity of *all* languages on this subject:[6]

it is unfortunate that most languages have no specific word for probability in its scientific sense but only popular terms like Wahrscheinlichkeit, probability, probabilité. However, no term has been invented and, naturally, it is quite possible for a scientific concept to exist without having a special name. This is the case with many of the most important concepts of mechanics which are hidden behind such ordinary words as force, mass, work, etc. All the same, I do feel that many laymen, and even some professionals in the field of mechanics, would understand these concepts more clearly if they had Latin names rather than names taken from everyday usage.

Thus Carnap is wrong if he thinks von Mises is unaware that he is changing the meaning of 'probability.' Furthermore, he is aware that (at least part of) the ordinary meaning which he is rejecting is the 'logical relation' type of probability, which Carnap associates with probability$_1$.[7]

It is brought out repeatedly in this book that the word 'probability' has a meaning in everyday language that is different from its quantitative meaning in probability calculus. Some authors with metaphysical leanings have sought to build a separate theory on this other meaning of the word. Such attempts, namely the study of questions of reliability or plausibility of judgments, of propositions and systems of propositions, are justified as long as they remain within certain limits. However, as soon as numerical values are attributed to these plausibilities and used in calculation, one has either to accept the frequency definition of probability (as is done by some authors) or fall back on an apriori standpoint based on equally likely cases (as is done by others). The stated purpose of these investigations is to create a theory of induction or 'inductive logic.' According to the basic viewpoint of this book, the theory of probability in its application to reality is itself an inductive science; its results and formulas cannot serve to found the inductive process as such, much less to provide numerical values for the plausibility of any other branch of inductive science, say the general theory of relativity.[8]

So von Mises has rejected the probability$_1$/a priori type of theory and forbidden the employment of probabilities in such logical uses as 'questions of reliability or plausibility.' What, then, remains in the scope of the concept?

Von Mises proposes to restrict probability theory to such special problems as games of chance, social mass phenomena (especially insurance and actuarial problems), and statistical treatments of

147

mechanical and physical phenomena. Such cases share a common feature which he thinks crucial; 'we find events repeating themselves again and again. They are mass phenomena or repetitive events.'[9] Individual or infrequent events, however, clearly do not possess this repetitive character, and are thus beyond the scope of probability theory.[10]

> We state here explicitly: The rational concept of probability, which is the only basis of probability calculus, applies only to problems in which either the same event repeats itself again and again, or a great number of uniform elements are involved at the same time. Using the language of physics, we may say that in order to apply the theory of probability we must have a practically unlimited sequence of uniform observations.

This restriction of the applicability of probability theory is formalized in the fundamental notion of a *collective*.

The term 'collective', von Mises says 'denotes a sequence of uniform events or processes which differ by certain observable attributes, say, colours, numbers, or anything else.'[11] His examples show that a collective must be large and he seems to require that it be based on some natural kind or similarity.

After thus defining (somewhat vaguely and ostensively, to be sure) the term 'collective,' he continues

> The principle which underlies the whole of our treatment of the probability problem is that a collective must exist before we begin to speak of probability. The definition of probability which we shall give is only concerned with 'the probability of encountering a certain attribute in a given collective....' It is essential for the theory of probability that experience has shown that in the game of dice, as in all the other mass phenomena which we have mentioned, the relative frequencies of certain attributes become more and more stable as the number of observations is increased...[12]
>
> Let us now add further precision to our previous definition of the collective. We will say that a collective is a mass phenomenon or a repetitive event, or, simply, a long sequence of observations for which there are sufficient reasons to believe that the relative frequency of the observed attribute would tend to a limit if the observations were indefinitely continued. This

limit will be called *the probability of the attribute considered within the given collective.*[13]

Von Mises adds one more restriction to his definition because of the existence of *regular* processes (the example he gives is a succession of large milestones and smaller 1/10-mile markers) which also have frequencies that tend to a limit after sufficient observation. These processes are rejected as being non-random. The concept of randomness in von Mises's theory requires that all relative frequencies in a random collection must remain the same if a subset is selected on the sole basis of place in the series. This independence of order is then included in the final definition of a proper collective:[14]

A collective appropriate for the application of the theory of probability must fulfill two conditions. First, the relative frequencies of the attribute must possess limiting values. Second, these limiting values must remain the same in all partial sequences which may be selected from the original one in any arbitrary way.

The final definition of probability is thus: The *probability* of an *attribute* is the *limit* of the *relative frequency* with which it appears in a *random collective.* Now that we have a definition of the fundamental concept, probability, and of its theoretical ground, the collective, development of the theory continues with the specification of its mathematical laws – the probability calculus.

Von Mises contends that all of probability calculus arises from four simple operations, just as all of algebra arises from addition, subtraction, multiplication and division. To specify or justify the calculus, therefore, one need only specify or justify these basic operations. They are:[15]

1 Selection – In which a partial sequence is derived from the original sequence by place selection. Attributes and probabilities remain the same in the new collective as in the old. (Example: if we only count dice throws immediately following an Ace, the probability of a Five is unchanged at 1/6.)
2 Mixing – Two or more attributes are 'mixed' (combined) into a single attribute whose probability is the sum of their individual probabilities. (The probability that a dice throw is either Three or an even number is $1/6 + 1/2 = 2/3$.)
3 Partition – In which a new collective is specified by a selec-

tion of attributes (rather than place numbers, as in Selection). The new frequencies are obtained by dividing each old probability by the sum of the probabilities of all selected attributes. (Example: The chance that an odd dice throw is a Three. The initial probability of a Three is 1/6. The sum of the probabilities of odd numbers is 1/2. $1/6 \div 1/2 = 1/3$.)

4 Combination – In which a new collective is formed by the simultaneous sampling of two independent collectives. The distribution of elements (ordered pairs) in the new collectives is obtained by multiplying the probabilities of each original element in the initial collectives. (The probability of a Five on a red die and an even number on a green is $1/6 \times 1/2 = 1/12$.)

We can now consider ourselves to be in possession of a full-blown theory, because we have a series of mathematical laws which establish the relations between the fundamental entities of the theory. (Not unexpectedly, those laws are substantially identical to the valid theorems derivable in a priori probability calculi.) Indeed, it is an *interpreted* theory, since its basic terms ('probability,' 'collective') have been linked to the real world by what Hans Reichenbach has called 'coordinating definitions.' This is the first formal development of the RF theory of probability.

Reichenbach's more fully developed, modern RF theory is based on a slightly modified version of von Mises' definition of probability:

> In order to develop the frequency interpretation we define probability as the *limit of a frequency* within an infinite sequence ...
> ... the *relative frequency* $F^n(A, B)$ is defined by
>
> $$F^n(A, B) = \frac{N^n(A \cdot B)}{N^n(A)}$$
>
> [i.e., the number of ordered pairs from a sequence which belong to the common class *A* & *B* divided by the number of sequence elements which belong to *A*].

With the help of the concept of relative frequency, the frequency interpretation of the concept of probability may be formulated:

> *If for a sequence pair $x_i y_i$ the relative frequency $F_n(A, B)$ goes toward a limit p for $n \to \infty$, the limit p is called the probability from A to B within the sequence pair.* In other words, the

following coordinative definition is introduced:

$$P(A, B) = \lim_{n \to \infty} F_n(A, B)$$

No further statement is required concerning the properties of probability sequences. In particular, randomness...need not be postulated.[16]

Thus Reichenbach's theory differs from von Mises's in that it purports to deal with all types of sequences (collectives) rather than merely random sequences. Reichenbach contends that von Mises's Axiom of Randomness is unnecessary and imposes an undesirable limitation of the scope of probability theory.[17] He further differs in considering both the concept of absolute probability and the probability of a single event to be meaningful and capable of interpretation by the frequency theory (see below).

Reichenbach agrees with Carnap that the statements of the probability calculus are merely tautological implications of the axioms and 'The question whether the individual probability statement is true or false, then, is not a problem of the calculus....'[18] Thus, in the use of the probability calculus:

> We assume the existence of some probability implications to be given; and we deal only with the question of how to derive new probability implications from the given ones. This operation exhausts the purpose of the probability calculus.[19]

Reichenbach, further, does not think (as von Mises claims to think) that the probability calculus is based on a generalization of past empirical data. He explicitly regards the calculus as an axiomatic mathematical system.

In his construction of the formalism of the probability calculus, Reichenbach employs four axioms.[20]

1 Axiom of univocality – Each probability implication has only one numerical value.
2 Axiom of normalization – Logical implications imply probability-implications of degree 1; all probabilities are non-negative numbers.
3 Axiom of addition for exclusive events.
4 Axiom of multiplication for relative probabilities and independent events.

151

Let us suppose that all standard probability laws can be derived from these axioms. Then if we wish to apply the calculus to some part of reality, we need only assume that the axioms are true of that part when they are interpreted by coordinating definitions. But Reichenbach has gone even further in reducing our assumptions – he has demonstrated that these axioms are true, and tautologically true, whenever the frequency interpretation is employed. That is to say, *if* symbols like '$P(A, B)$' are replaced in the axioms by phrases like 'the limit of the relative frequency of A's which are also B's as the number of A's increases without bound,' *then* the axioms become relatively simple arithmetical statements about the properties of limits in infinite sequences. Those statements are such that they can be shown to be true using only the axioms required for arithmetic and deductive logic.

Here Reichenbach has done consciously what von Mises did unwittingly; he has deduced the probability calculus from his definitions of 'probability,' 'relative frequency,' 'sequence,' and 'limit.' The result is not, as von Mises thought, that the laws of the probability calculus have been shown to be supported by empirical generalization, but rather, as Reichenbach asserted, that the calculus is shown to be applicable wherever the frequency interpretation applies to reality. Or, in other words, we can know that every statement licensed by the probability calculus is certainly true *if* we only know that every 'P' refers to an existing limit of an infinite sequence. The rest is simple mathematics.

The definition of '$P(A, B)$' in terms of the limit of a relative frequency has therefore greatly simplified the metaphysical interpretation and mathematical treatment of probability statements *if* it is strictly adhered to. That is, we can plainly understand (and usually agree with) someone who says 'The limit of the relative frequency of A's which are B's is p and the limit with which A's are C's is q; since no B's are C's, I conclude that the limit with which A's are either B's or C's, is equal to the sum of p and q.' If the same statement is changed from the assertoric to the hypothetical mode ('If the limit...then I conclude...'), we can *always* agree with it, since it merely express a mathematical truism. Now this is clearly *not all* that any speaker ever intends to assert by a probability statement. Even worse – it is not all that Reichenbach intends to assert by it! He has given us, for example, *another* definition of '$P(A, B)$' in *The*

Theory of Probability, when he first sets out to explain what probability statements are.[21]

> Probability statements... have the character of an implication; they contain a first term and a second term, and the relation of probability is asserted to hold between these terms. This relation may be called probability implication. It is represented by the symbol
>
> $$\underset{P}{\supset}$$
>
> This is the only new symbol that the probability calculus adds to the symbols of the calculus of logic. Its connection with *logical implication* is indicated by the form of the symbol: a bar is drawn across the sign of logical implication. Whereas the logical implication corresponds to statements of the kind, 'If *a* is true, then *b* is true,' the probability implication expresses statements of the kind, 'If *a* is true, then *b* is probable to the degree of *p*.'... *The probability statement is a general implication between statements concerning a class membership of the elements of certain given sequences.*

Finally, the '*P*-notation' is introduced so that (with certain restrictions concerning the empty class as reference class) we can talk about the numerical value, *p*, by writing '$P(A, B) = p$.'

Now this definition of '$P(A, B)$' in terms of a 'probability implication' is quite different from that in terms of frequency limits. It is clearly *intended* merely to relate to symbol '$P(A, B)$' to the word 'probable' or 'probability' so that the subsequent definition of '$P(A, B)$' will be part of a theory of probability, but *in fact* it says already something about the meaning of 'probability.' Furthermore, what it says is, in general, false.

I refer the reader to what Carnap said about formulating probability statements in the 'if – then' form (*Logical Foundations of Probability*, pp. 30–33). Briefly, the point is this: Assume $P(A, B) = p$. Let '*A*' be 'is consumptive' and '*B*' be 'will die within the year.' Let '*A*' be true of Willis Ray Surrett. Translating '$P(A, B) = p$' into a probability implication, we get the general form:[22]

$$(X)[A(X) \underset{p}{\supset} B(X)].$$

By Universal Instantiation we get

A(Willis Ray Surrett)$\Rightarrow$ B(Willis Ray Surrett).

According to what Reichenbach has said, this means

If Willis Ray Surrett is consumptive, then the probability that he will die within the year is p.

Now Sgt Surrett is a friend of mine, and I happen to know that he is in fact consumptive. Therefore, we use *modus ponens*, $[(p \supset q) \cdot p] \supset q$, to conclude: 'The probability that Willis Ray Surrett will die within the year is p' where 'p' will subsequently be defined as the limit of the relative frequency with which consumptives die within the year.

Now the conclusion is in fact false, since Ray Surrett is a young and quite healthy chap who happens to have a very mild and easily controlled case of tuberculosis. His chances of surviving the year are far better than those of consumptives in general. What has gone wrong here?

The problem is in the attempt to define probability relations in terms of implications. The essence of an implication is that once the antecedent is true it can be dropped as unnecessary and the consequent can be asserted alone (either categorically or, as Reichenbach thinks, as probably true). But Carnap and Reichenbach agree that no probability can be stated alone – they must always be relative to some reference class. *Therefore the relation between membership in the reference class and membership in the quaesitum class cannot be any form of implication.*

Now the point of all this lengthy excursus is not just that Reichenbach has made a quite common error in trying to explain the probability relation. The point is that this error arose when he tried to tie the clearly defined mathematical symbol '$P(A, B)$' to the word 'probability' and its use in ordinary talk about probability relations. In all that follows, therefore, we must be constantly aware of the following points arising from Reichenbach's definition of probability:

1 It is unquestionably true that the limit S of rf's in infinite sequences obey the mathematical laws normally included in the probability calculus.
2 Therefore there is no logical or mathematical problem in the manipulation and transformation of these limit values according to those laws.

3 *Calling* these values 'probabilities' raises no problems about these values or their manipulations.

4 Saying that all probabilities are limits of frequencies is quite a different matter and raises such points as

  (a) Probabilities are frequently known; are limits?

  (b) Some sequences have no limits; do some events have no probability?

  (c) What meaning applies to the probability of single events?

  (d) Does 'probability' *always* mean 'limit of relative frequency?'

5 We must therefore be very careful to determine when Reichenbach is unquestionably talking about uncontroversial mathematical truths and when he is talking about controversial interpretations of words like 'probability.'

3   SOURCE OF INITIAL PROBABILITIES

The theory of probability as a limit of the relative frequency of occurrence outlined in the previous section, which we have attributed to von Mises, stands in rather surprising contrast to his repeated assertion that it is no part of the probability theorist's job to determine initial probability values.[23] That function is left entirely to the statistician.

Instead, von Mises seems to believe[24] that all probability theory can do is to derive new probabilities from old ones:

> A great number of popular and more or less serious objections to the theory of probability disappear at once when we recognize that the exclusive purpose of this theory is to determine, from the given probabilities in a number of initial collectives, the probabilities in a new collective derived from the initial ones.... A probability can only be determined from the knowledge of other probabilities on which it depends.

Now the derivation of new probabilities from old ones is the task of the calculus of probability – a well-known and generally agreed upon branch of pure mathematics. It is true that there are alternative formalizations of the calculus, using different axioms and notations, but each of these includes the same basic notions and generates the same set of fundamental theorems.

And while it is certainly true that there remain problems to be solved in the development of this branch of pure mathematics, von Mises's book is not famous for its solutions to such problems. Quite the contrary. While von Mises did do some original and widely recognized work on the mathematics of indefinite series, his place in probability theory, and especially the fame of *Probability, Statistics, and Truth,* are secured by what he said about the *interpretation,* the *meaning,* the *application,* and the *justification* of probability theory – in short, his greatest achievement was his definition of 'probability' and his explanation of the source of initial probabilities. Odd that a man's fame should rest on something which he denied he was doing!

The fact is that, despite his disclaimers, von Mises *did* tell us where to look for initial probabilities and did so in a way that differed from traditional theories and started a new version of probability theory. He abandoned equiprobable cases in favor of the relative frequency in an empirical series. His apparent downgrading of the role of initial probabilities is, I think, an accident of history and biography which happened somewhat as follows.

Von Mises was closely associated with the Vienna Circle during the exciting years of the early development of scientific empiricism. He was no doubt affected by the intense debate about the nature and function of scientific theories. As a mathematician himself, he presumably took great interest in Reichenbach's seminal works on the nature of geometry.[25] In fact, I submit that *von Mises probably thought he was doing for probability theory what Reichenbach had done for geometry.* A strict parellelism between the two would require von Mises to exhibit probability theory as a pure mathematical theory whose validity is independent of *any* contact with the real world. He could never bring himself to do this, since he *also* wanted to think of probability theory as an inductive science, firmly grounded on empirical facts.[26] But he could and did adopt the notion that the pure theorist is unconcerned with the source of his data. In fact he does so by explicit analogy to the role of the geometer:[27]

The geometer does not ask how the lengths of the two sides and the magnitude of the angle have been measured; the source from which these initial elements of the problem are taken lies outside the scope of the geometrical problem itself. This may be the business of the surveyor, who, in his turn, may have to use many geometrical considerations in his work. We shall find an exact analogy to these relations in the interdependence of statistics and probability.

If von Mises had strictly followed his own analogy, he would have produced only a relatively undistinguished axiomatization of the probability calculus. Fortunately for us, he did not merely abandon to the statisticians the task of obtaining initial probabilities, he told them how to do it. Specifically, he told them *not* to count equiprobable cases, etc., but instead to obtain the limit of a relative frequency in a random empirical sequence. Reichenbach and all other RF theorists agree that this is the way to establish initial probabilities.

This directive to statisticians is of fundamental importance. It abolishes at one stroke every probability theory (practically) from Pascal to Keynes. But it also poses a new and perplexing problem: How does one establish the limiting value of a relative frequency in an infinite empirical sequence?

In Dice Games

RF theorists generally reject the methods and explanations developed in the CTP and the AP theory of probability.

Against the Principle of Indifference, von Mises has argued that when we are ignorant we can say nothing about probabilities[28] and when we can give probabilities it is because we have knowledge of some frequency.[29]

Against the general idea of a logical theory of probability, Reichenbach declared that 'Logic cannot supply a probability metric; only experience and observation can inform us about degrees of probability or about the equality of such degrees.'[30] And again, '*there is no a priori ascertainment of a degree of probability; a probability metric can be determined only a posteriori.*'[31]

It follows that our method of establishing initial probabilities in dice games must depart from the traditional method of counting equipossible cases and a new procedure must be established. For von Mises, that procedure is quite straightforward: initial probabilities in dice games are established by counting relative frequencies. The traditional, equal, values are well established by historical observation and may normally be used without extensive counting. If 'unusual' results occur repeatedly in some game it is prudent to treat it as a special case which requires the establishing of an initial probability by counting a frequency. Even biased dice are predictable by this method.

Reichenbach offers a more detailed analysis:[32]

> There are three possible ways for *a posteriori* establishment of a probability metric:
>
> 1 Degrees of probability can be directly ascertained through induction by enumeration (statistical probabilities).
> 2 A probability metric can be inferred deductively from known probabilities (deduced probabilities).
> 3 A probability metric can be inferred by means of general inductive methods from known observational data (hypothetical probabilities).

These three methods, however, are not equally applicable in every situation. 'For primitive knowledge – a state of knowledge that does not include a knowledge of probabilities – the rule of induction is the only instrument for the ascertainment of probabilities.'[33]

In *advanced knowledge* probabilities can also be ascertained by methods 2 and 3 above. But first we shall look at the basic method for establishing initial probabilities, the Rule of Induction (by enumeration).[34]

> RULE OF INDUCTION. If an initial section of n elements of a sequence x_i is given, resulting in the frequency f^n, and if, furthermore, nothing is known about the probability of the second level for the occurrence of a certain limit p, we posit that the frequency $f^i(i > n)$ will approach a limit p within $f^n + \delta$ when the sequence is continued.

To apply this rule to our example of a dice game we must first decide what we wish to take as the reference class (throws of this die) and what is to be the quaesitum property (Five uppermost). Then we start throwing the die and counting the Fives. At any time thereafter, we posit[35] that the ultimate limit of the frequency (which is the same as the probability) will be equal to the heretofore observed relative frequency (within a small but unspecified range of uncertainty δ).

This rule has come to be known as the Straight Rule for Induction, because it is so straightforward and because it implies that any graph of the future values of the relative frequency will be a straight (and level) line. As formulated by Reichenbach, it has no lower limit on the number of required observations. It follows, therefore, that after the first throw has been observed we must conclude that the probability of a Five in the long run is equal to 1 or 0, depending on whether that throw was or was not a Five. This strikes many people (including Carnap) as being counter-intuitive. Likewise, after the second throw, we will be abjured to act as if the probability of a Five is 1, or 1/2, or 0, depending on the first two results. But as the number of observed throws, N, increases, it becomes less and less counter-intuitive to identify the probability with the observed frequency value. After 100 throws, for example, with 18 recorded Fives, we might still be inclined to think that the probability is 'really' 1/6 and that 0.18 is only an approximation which is slightly off because of the smallness of the sample. But after 10,000 throws, if Five has occurred 4,999 times, even the most hidebound a priorist will be willing to admit that this is not a 'normal' die and that the 'best estimate' of the probability is 'around one-half.' Perhaps the chief practical problem in establishing initial probabilities by the Rule of Induction is deciding how many observations is 'enough.'

Fortunately for those of us who like to gamble, it is no longer necessary to sit down and record hundreds of dice throws before getting into a crap game. The reason is, of course, that we are no longer in a state of 'primitive knowledge.' We are therefore entitled to use the last two of Reichenbach's methods in determining probabilities.

As an example of the method of 'deduced probabilities,' suppose we know that each of two dice has a 1/6 probability of showing a Five. By a straightforward application of the probability calculus we may deduce that the probability of a Double Five is equal to 1/36.

For 'hypothetical probabilities,' suppose a player in our crap game requests a new pair of dice. From our previous experience with other dice (and with the honesty of the gambling establishment), it is reasonable for us to infer that the probability of a Five on one of the new dice is also 1/6. This is clearly not a deductively valid conclusion, but has the slightly problematic and generally reliable nature of other well-founded inductive conclusions.

These two methods, as illustrated, can give us probability values to act on, but they are not, strictly speaking, initial probabilities. In each case, we must already know the value of some probabilities before we can apply the methods. We conclude, therefore, that the only source of initial probabilities in Reichenbach's system is the application of the Rule of Induction to observed empirical sequences.

IN ACTUARIAL CASES

By actuarial case we mean one in which a fairly sizeable body of statistical information exists. This includes the frequencies and data of the physical sciences, as well as the more traditional actuarial examples from insurance and the social sciences. Such cases constitute a paradigm case for RF theory because:

1 They exhibit stable relative frequencies which have been found to be excellent guides to future occurrences.
2 No other established method works nearly so well in these important cases.

If we wish to know probabilities of death or marriage, or to predict rates of radioactive decay or distributions of stellar magnitudes, the proper, indeed the only satisfactory, way of proceeding, according to the RF theorists is to perform a straightforward inference from an observed relative frequency to a probability of equal numerical value. The process is as follows:

First, identify the collective (von Mises) or reference class (Reichenbach). Remember that the collective of 40-year-old men and the collective of 40-year-old men in New York are not identical and may exhibit properties with different frequencies. Decide which group you are really concerned with.

Second, look up the relative frequency of the property among recorded members of the collective. If 40,000 individuals are in the records and 4,000 of them have the property, the relative frequency is 0.10.

Third, check to see if this value is stable over time and subgroups. (Here von Mises shares Keynes's respect for Lexis's tests for significant dispersion.)

Fourth, accept any stable value as the limit of the relative frequency of occurrence – and hence the probability – of the property in the collective.

This is basically the Straight Rule, hedged around with a few safeguards. By requiring large samples and testing for randomness and super- or sub-normal dispersion von Mises hopes that we can escape many of the difficulties of that rule. But his and Reichenbach's fundamental inductive philosophy is the same: *observed regularities in experience will continue with present frequency values.*

This method has worked for insurance companies for years, and would have worked for gamblers had they not preferred the simpler method of uniform distribution. It represents perhaps the greatest advantage of the RF school over the Classical – it allows us to generate initial probabilities in cases where no equiprobable alternatives exist.

In summary, then, initial probabilities in actuarial cases are established by simply counting the relative frequency of the property in the reference class and accepting that as the limiting value of the frequency, which is also the probability.

4   THE PROBABILITY OF A SINGLE EVENT

According to von Mises, there is no such thing as the probability of a single event. For example,[36]

> When we speak of the 'probability of death,' the exact meaning of this expression can be defined in the following way only. We must not think of an individual, but of a certain class as a whole, e.g., 'all insured men forty-one years old living in a given country and not engaged in certain dangerous occupations.' A probability of death is attached to the class of men or to another class that can be defined in a similar way. We can say nothing about the probability of death of an individual even if we know his condition of life and health in detail. The phrase 'probability of death,' when it refers to a single person, has no meaning at all for us.

This is the obvious, and the usual, position of most RF theorists, but Reichenbach is different because he thinks the RF theory is applicable even to single events. According to him, the frequency interpretation deals with individual cases by means of a *posit*. Rather than asserting the most probable alternative as true, we *posit* it by behaving as though it were true. '*A posit is a statement with which we deal as true, although the truth value is unknown.*'

In order to act rationally, we should always make those posits which would be most successful if repeated infinitely. Thus we should posit that alternative which would appear with the greatest relative frequency in an extended sequence, which is the same as to say that we should always act on the most probable alternative (in the RF sense). In this way a statement about an individual probability is seen as an 'elliptic mode of speech' which acquires a 'fictitious meaning' by a transfer of meaning from the general to the particular case. The adoption of the fictitious meaning is justifiable, not for cognitive reasons, but 'because it serves the purpose of action to deal with such statements as meaningful.'[37]

What is most important about Reichenbach's view is that it involves the introduction of a new sense of the term 'probability.' We cannot avoid von Mises' contention that, if 'probability' is defined as 'limit of relative frequency in a series,' it just *is* meaningless to speak of the probability of an individual event. Reichenbach is well aware of this, but he is also committed to making the frequency interpretation do for all uses of 'probability.' The result is his insistence on such terms as 'elliptic,' 'derivative,' and 'fictitious' in order to explain how such statements are meaningful in a different way but don't have a different meaning. This seems to be an equivocation. If the term 'probability' has a meaning in statements of individual events and if that meaning is not 'relative frequency' then it is a *different* meaning and Reichenbach should say so.

That Smith will roll a Five

The fictitious transfer of meaning from the sequence to the individual, which Reichenbach believes sufficient to give 'meaning' to individual probability statements, does not proceed from a sequence of events to a single event but from a sequence of propositions to a single proposition. This involves a shift in the way of conceiving probability, from a property of (sequences of) events to a property of (sequences of) propositions. Reichenbach credits George Boole with first recognizing the possibility of this alternative interpretation,[38] but the blame is entirely Reichenbach's for choosing to call it the 'logical interpretation of probability' in contrast to the 'object interpretation of probability.'[39] Perhaps 'blame' is too harsh a word, since in Reichenbach's time there did not exist the present extensive use of 'logical' to refer to the type of probability theory which I call 'a priori' (although Karl Popper spoke of 'the logical interpretation of probability' in the modern sense in his *Logik der Forschung*[40]

162

published the very same year as *The Theory of Probability*). Still, it is certainly confusing *now*, since the phrase has acquired a fixed meaning different from Reichenbach's. To avoid this confusion, I will use 'propositional interpretation' where Reichenbach uses 'logical interpretation' to show that the things which are probable are linguistic rather than physical.

Briefly, the propositional interpretation identifies 'probability' with 'truth-frequency in a sequence of proposition.' In our present example, one sequence of propositions hz_i ('z_i is a throw of this die') is determined as true or false by a sequence of events x_i which either are or are not throws of a die. Another sequence, fx_i ('This is a Five') is determined as true or false by a sequence of events which either are or are not groupings of Five pips. The frequency interpretation is constructed by counting the number of true propositions 'fx_i' within the sub-sequence selected by true propositions 'hz_i.'[41] Thus if we say of each throw 'This is a Five' we shall have a sequence of propositions, some of which have 'true' and some 'false' as their truth-value. The probability of 'This is a Five' is equal to the limit of the relative frequency with which 'true' occurs in that sequence. So far the propositional interpretation is isomorphic to the object interpretation.

Now if we consider the individual proposition 'This is a Five' corresponding to the single event which is Smith's next roll of the die, we can maintain that proposition 'only in the sense of a posit'[42] since we do not yet know if it is true or false, but we can transfer to that posit the probability value obtained for the sequence as a whole. That value is then said to be the *weight* of the posit: 'the probability of the single case, therefore, is regarded, in the [propositional] interpretation, as the weight of a posit. A posit the weight of which is known is called an *appraised posit*.'[43]

The concept of appraised posits is further elaborated by Reichenbach into an entire system of multi-valued probability logic. The development of this system is an intriguing intellectual exercise. It starts with a formalistic conception of logic as a whole and models itself somewhat after modal logic. It issues in a set of 'truth tables' which specify the probability values of combinations of statement sequences when the probability values of the sequences themselves are given. This logic of statement sequences is then transformed into a 'fictitious transfer of the truth properties of propositional sequences to individual propositions.'[44]

163

But despite the impressiveness and logical rigor of Reichenbach's systematic development of his logic, it is easy to see that the basic task is the transformation of the probability calculus from a system dealing with properties of objects to a system dealing with properties of propositions. His system differs essentially from Carnap's and Keynes's only in the assertion that the probability of a propositional sequence is defined by the frequency of truth of the propositions within the sequence and the probability of an individual statement is a fictitious transfer of the truth properties (i.e., probability) of some sequence.

A peculiar characteristic of Reichenbach's system is that although he starts by replacing the traditional, alternative truth-values of 'true' and 'false' by the infinitely divisible scale of probabilities from 0 to 1, he ends up with a system in which the only allowable assertions are those with a probability of 1. By using a device called 'quantitative negation' it is possible to assert the equivalent of 'A has a probability of p' and to do so in the form of a statement which itself has a probability of 1. But this differs from asserting the probability-statement *as true* only in the same way that 'true' and 'having a probability of 1' differ.

Carnap has argued[45] that Reichenbach's attempt to replace truth values with probability values is misguided because

> these views are based on a lack of distinction between 'true' on the one hand, and 'known to be true', 'absolutely certain', 'completely verified', 'confirmed to the maximum degree', 'having the probability$_1$ 1', on the other. The concept expressed by the latter phrases in their strictest sense is indeed an absolutistic concept that should be replaced by the concept of probability with its continuous scale of degrees. Both these concepts refer to given evidence; the concept of truth, however, does not and this is seen to be of an entirely different nature, and, hence, values of probability$_1$ are fundamentally different from truth-values.

Since Reichenbach only allows assertion of sentences with probability 1 and uses only *modus ponens* for his rule of inference, it is hard to see how its logical scope could differ from Carnap's system where all probability statements are true (analytic) and *modus ponens* is the rule of inference. It is only the attempt to *interpret* the system in terms of relative frequencies of truth which is different.

I agree with Carnap's criticism and conclude that the system of 'many valued probability logic' rests on a confusion and is less radical an innovation than Reichenbach thought. For von Mises, of course, the entire discussion is pointless, since there is no such thing as the probability of individual events.

That Smith will be elected Mayor

I have reversed the order of presentation of our running examples of single events because, for Reichenbach at least, the probability of being elected mayor depends on the theoretical analysis used for dice games – and then goes on to difficulties of its own. Therefore, it is convenient to begin this discussion by accepting Reichenbach's explanation of how a single event like the throw of a die receives a probability in his theory. Then we must try to apply this method to Smith's chances of being elected mayor.

The fundamental difficulty in problems of this type is the identification of suitable reference and quaesitum classes. This is commonly called the 'problem of the reference class.'

Suppose we start with the most general possible reference class, inhabitants of this city. Then our probability seems to be something like 1/50,000, since each election chooses only one of the 50,000 inhabitants of the city. By adding the information that Smith is male, we can reduce the odds to something like 1/30,000, if his city follows the American chauvinist tradition of rarely electing female mayors. But now we run into a difficulty: should our fudge factor for sex be based on the historical rate of election of women in this city, or the historical rate for the country as a whole or the near-present rate of one or the other? Reichenbach's rule of thumb – select the narrowest class for which reliable statistics exist – would seem to indicate that statistics for this individual city would be preferable. But, as it happens, the US is currently experiencing a major increase in the political activity (and success) of women. This factor should certainly be included in our calculations – perhaps by using near-present statistics for the country as a whole. Even so, it is unlikely that any numerical factor directly based on past experience could adequately reflect the extent of the improvement in a woman's chance of being elected now or in the future. Such statistics inevitably lag behind reality – the difficulty is to identify those which lag behind

165

the least. Difficulties of this kind apply to the selection of all statistical factors (Republicans may be increasing in the South while decreasing nationally – which data do we use for Louisville?).

It may be that in some cases we have *no* reliable statistics which are obviously relevant. If Smith is, for example, a Liechtensteinian immigrant who was converted to Buddhism and who is running against a red-headed, left-handed mushroom farmer who has changed parties twice, it is difficult to see how we could appeal to statistics for a reliable prediction. Some single events may be so qualitatively rare that there are no important empirical series which can be used as predictors.

A final difficulty in Reichenbach's theory is that for some possible single events *there may not be a probability at all.*[46]

> When a die is thrown upon a table, it is possible that a sudden thunderbolt may happen simultaneously; but such a statement of possibility does not mean that a probability implication exists between the two events. I do not wish to say that the probability is very small; I mean, rather, that it is not permissible to assert a definite regularity with respect to the occurrence of thunder when the die is thrown repeatedly. The illustration will make it clear that the existence of a probability cannot be inferred from the possibility of an event.

'The existence of a probability,' in Reichenbach's system, is completely dependent upon the existence of a limit of a relative frequency in an infinite series. He thinks this is a reasonable assumption, in many cases, because investigation has revealed a striking regularity in many empirical series. But even Reichenbach does not think *all* empirical series must tend to a limit (we know some mathematical series do not). At least some identifiable series may have no limits – from which it follows that some (possible) empirical events may have no (defined) probability.

The reader may find it strange to think that there is *no* probability that this throw of the die may be accompanied by a thunderbolt (remember this is not the same as saying there is a *zero* probability), but he should remember that this is not unique to Reichenbach's system, nor is it a necessarily fatal flaw. It is true that Carnap would argue for the existence of a determinate (on given evidence) and in principle computable probability in this case, but Keynes would say that, though real, the probability is indeterminate. And von Mises would deny that *any* single event has a probability. So the *mere* failure to specify a determinate value for the probability of a lightning bolt accompanying a dice throw need not be a defect in a probability theory. But Reichenbach is not just aiming at *a* probability theory. He is trying to convince us that his RF conception of probability is completely adequate to capture *all* uses of the concept *probability*. If our intuitions tell us that every possible event does have some probability, then we must conclude that Reichenbach has failed in his aim.

5 PROBABILITY OF REPETITIVE KINDS OF EVENTS

Here we have the heart of RF theory. A relative frequency is essentially a property of a repeatable kind of event, and all of the unqualified successes of RF probability theory have dealt with such kinds of events. We will find here, or nowhere, the reasons for preferring RF theories.

That a thirty-year-old Man will get Married

This case can be taken as a paradigm of RF theory. If we keep in mind that we are referring to a (some) thirty-year-old man, or the class of thirty-year-old men, but definitely not a particular, individual man, we can see that this example has the following favorable features:

1 There exists a known, stable frequency of occurrence in the class. By checking the marriage records throughout the US we can establish that of every 1,000 men single at their thirtieth birthday, 50 (say) marry before their thirty-first.
2 The property is random in the class. If we pick out every 7th such man, we still find 5 per cent marrying during the year. The same is true for those whose last name starts with a 'C' or an 'L.' Within acceptable limits, it is also invariant with geographical location. (This randomness or insensitivity to place selections is necessary to make the class suitable as a collective for von Mises.)
3 The property is suitable for treatment by the probability calculus. For example, the probability that such a man either

167

marries or does not marry is $0.05 + 0.95 = 1$ (Unit Sum). The probability that he marries *or* has a name beginning with '*L*' is (say) $0.05 + 0.02 = 0.07$ (Addition Rule), while the probability that he marries *and* has a name beginning with '*L*' is $0.05 \times 0.02 = 0.0010$ (Multiplication Rule).

4 There is sound empirical evidence for the probability state-ment '$P(M_{30}) = 0.05$.' This evidence consists of marriage records, personal observations, etc. The frequency has remained stable for years (we will suppose) and no sub-group has unexpectedly varied from the relative frequency 0.05.

5 The probability statement '$P(M_{30}) = 0.05$' leads (and has led) to success in predicting and controlling the future. Anyone who doubts the value of such statements should try to determine the total assets of insurance companies in America. This astonishing sum has been accumulated through skillful and successful prediction of the behaviour of large groups of people. Although insurance companies might have no interest in this specific example, governmental bureaucrats, demo-graphers, and sellers of wedding apparel will.

6 The example admits of no interpretation in terms of equi-probable alternatives. Classical theory could not deal with it at all. Keynes could only deal with it by invoking an intuitive probability that the observed frequency will continue, and this intuitive probability, since it is not based on the principle of indifference, could not be numerically precise. The ability to deal with such cases is, I repeat, the greatest virtue of the RF theory of probability.

Now it would seem that this example, or others which share most of these six characteristics, would be perfectly suited for the applica-tion of RF theory. Unfortunately, the case is not as simple as it appears.

Let us suppose, to exhibit this problem, that the probability of a 30-year-old man's getting married is 0.05 and that we know that it is. We have therefore waived all metaphysical and epistemological difficulties about the existence of a limiting value and our ability to discern that value.

We are now in possession of a known-to-be-true probability statement, '$P(M_{30}) = 0.05$.' *Now that we have it, what can we do with it?*

Von Mises has explicitly forbidden us to use it to assess the likelihood that a thirty-year-old like Jack Warndof will get married this year. Such attributions of probability of individual events is meaningless and not to be condoned under any circumstance.[47]

Clearly, then, what von Mises intends us to do is to apply it to groups. Suppose we say that we can predict, using the theory of probability, that approximately five of the 100 thirty-year-old bachelors in Jack's Law School class will get married this year. Unfortunately, this is also not acceptable to von Mises. All that the probability statement *says* is that a certain proportion of the infinite collective of thirty-year-olds has the property of getting married. This has no more implication for any small group of thirty-year-olds than it has for any individual. The proportion of 0.05 marriages *on the whole* is perfectly compatible with any value whatsoever for this group of 100. Therefore we can say nothing about the composition of Jack's law class.

At this point it is traditional for theorists from Poisson on to invoke the Bernoulli-Poisson Theorem, or the Law of Large Numbers as it is often called. We explained this theorem, together with von Mises's criticism of it, in the section on chief criticisms of Classical theories.[48] One of the points von Mises repeatedly makes is that neither his theory nor the Law of Large Numbers says anything about the composition of any small group.[49]

It is essential to remember that the theory predicts not the exact result of a single sequence of observations but the outcome of the great majority of identical experiments (each experiment consisting of a large sequence of observations), repeated a very large number of times.

This means, of course, that nothing is to be gained here by invoking the Law of Large Numbers, since 100 is far too small a number. Again we must increase the size of the group with which we are dealing.

Let us next take the group of all thirty-year-old bachelors in America. Surely we can say that approximately 5 per cent of them will get married this year? Well, not quite.

If the group is large enough (and we shall finally assume that it is), von Mises's theory and the Law of Large Numbers imply that *there is a very high probability* that 5 per cent will get married next year. This is the basis of the success of insurance companies. If an

169

insurance company insures groups of 100 on the assumption that 5 members of the group will get married, and if the number of groups gets very large, then there is a very high probability that the assumption will be vindicated by an actual frequency of occurrence which is very close to 0.05. The import of Bernoulli's Theorem is that this 'probability of vindication' can be brought arbitrarily close to 1 by continually increasing the number of groups insured.

But a probability of vindication (like a probability of occurrence) gives us no license, on von Mises's view, to think that we will be vindicated (x will occur) *this time*. All we can say is that if the insurance company repeats this massive experiment an infinite number of times, then a very high proportion of those repetitions will be successful. We cannot say anything about *this particular* large set of insurance policies. So even insurance companies are disbarred from actually applying RF theory. What has gone wrong?

We started out trying to apply a probability statement to an individual. As expected, von Mises told us that wouldn't work. The same result applied to small groups of individuals. But von Mises and the Law of Large Numbers seemed to promise us that we could apply probability directly to large groups of events.

That turned out to be not quite the case; instead, we had to deal with large numbers of repetitions of large groups of events. Even at this level, all we can say is that it is 'very probable' or 'almost always occurs' that such a hughmongous group has a certain characteristic. It should be obvious by now that an infinite regress occurs whenever we try to apply the theory. Each successive application of the Law of Large Numbers leads only to another (very high) probability. But a probability statement by definition applies only to an infinite sequence and says nothing about the composition of any finite sub-sequence.[50] Therefore we arrive at our first proposition:

(1) No RF probability statement, strictly speaking, says *any-thing* about *any* finite event, group of events, or series.

This is obviously not a satisfactory way of looking at things. Our interest in the infinite series of 30-year-olds is not primarily a theoretical concern about the nature of infinite series, it is a practical concern about the fate of real, finite groups of people. A probability theory which gave us no help at all in dealing with our finite world would have few adherents. Therefore we must reduce the stringency of our interpretation, in order to make some application possible.

One way of trying to get around this problem is to say that if the limit of the frequency is 0.05, and if we act as if 5 per cent of real, finite thirty-year-olds get married, we will be right in the long run. This is a characteristically Reichenbachian thesis, which we will discuss later – the two best replies to it are, first: if 'the long run' refers to any finite length of time, it is false that we have any guarantee of success, and second: if it refers to an infinitely long run, then, as Keynes said, 'in the long run we shall all be dead.'

Von Mises more typically ignores the problem rather than trying to solve it. He sometimes slips insensibly from 'it is very probable that' to 'the majority will be.' He is always ready to make the illicit transition from RF probability to the 'likelihood of occurrence' which he had originally tried to dispense with – or worse, to a certainty of occurrence which is unjustified on any (standard) theory.

But our intention was to relax the stringency of our interpretation. Aren't most of us actually willing to agree that the Law of Large Numbers works very well in practice and that certain distributions of repetitive events really are very likely in the long run? Suppose we agree, then, that the theory can be applied to large but finite numbers of repetitive events. Let us agree, for example, that if we take all American thirty-year-olds in random groups of 1,000, it is very probable that the vast majority of these groups will have approximately 50 who get married. We further agree that it is reasonable to act on this probability (after all, that's how insurance companies get rich).

Now consider the event, E_t, which consists of the marriage-distribution of thirty-year-old American males at the present time, and the property, S, of being approximately equal to the distribution of the previous paragraph. We argued earlier that the strict meaning of the RF probability-statement above is that in an infinite sequence of E_t's almost all have the property S. Our current stipulation amounts to saying that this high probability in the long run also applies to the single event E_t. It is very probable that E_t has the property S and it is rational to act on the probability. But E_t's being S is a single event out of an infinite sequence – it's true that E_t is very complex and embraces a huge number of individual elements, but it is *in principle* no different from getting a seven on the next pair of dice (perhaps getting a sum greater than 101 on a roll of 100 dice would be a better example, since the probability of E_t being S is quite high).

171

From all of this we come up with our second proposition:

(2) If RF theory is applicable to any finite group of events, it is applicable to single events.

Originally we expected to agree with von Mises that RF probability finds its principal (only?) application in cases like the marriage prospects of thirty-year-olds as a class. Our careful analysis of the logic of the matter, however, leads to the surprising conclusion that the theory is *strictly* not applicable to any finite group, and if the stringency is relaxed it is applicable to single events as well.

That a Dice Throw will be Five

RF theorists believe that probability in a dice game is exactly the same as every other manifestation of probability – it is a property occurring in a random infinite sequence with a relative frequency that tends towards a limiting value. It is pragmatically valuable and interesting but of no theoretical importance that a well-made die has the same limiting frequency of occurrence for each of its faces. Clearly this is so because fairness is the principal design criterion for (most) dice, not because it has six possible outcomes of which we are equally ignorant.

The 'fair die' was prominent in the development of the Classical theory, and continued to be of some importance in theories of a priori (AP) probability. With von Mises, however, the unfair, biased, or loaded die comes into its own.

It was a minor scandal of earlier theories that they purported to predict future events with accuracy so long as we were ignorant or had only symmetrical information about the alternatives, but as soon as we learned that one alternative was more likely than the others, the old theories became *unable* to say anything about the future. This leads to the paradox that if we wish to predict the future we do better *not* to gather information, lest we violate the conditions of the theory.

No empiricist could stand for any such attitude; and of course beginning to engage in a game will start the flow of information, so we know people will not remain for long in a state of 'equal ignorance.' If they pass from that state to a state of 'equal knowledge,' all remains well. The Principle of Indifference can be employed with even greater confidence if we have positive experience confirming the equiprobability of the alternatives. But if our newly acquired knowledge begins to indicate that one dice face is more common than the others, what are we to do?

Some early theorists tried to construct alternative models, retaining some form of equiprobability. We might, for example, identify three equiprobable alternatives: the first is a Five, the second is a Two or Three, and the third is a One, Four, or Six. Besides being arbitrary and *ad hoc*, such models work even moderately well only in cases with fairly regular frequency distribution – they certainly are not a satisfactory general solution to the problem.

Von Mises, however, could argue that his theory *did* constitute a general solution, since it could satisfactorily deal with both fair and loaded dice, with no special assumptions or *ad hoc* models required. All one has to do is collect evidence by observing the die and then project the relative frequency of each side's occurrence into the future. This will give one usable probability values whether the die is loaded or not.

Of course the same application problems are found in the dice game as in the above sections. We still are told that probability statements apply only to very large samples and never to single events. This may seem reasonable as a partial explanation of why large casinos can count on a regular income but individual gamblers cannot. It seems strange or even ludicrous, however, if it means that a man who wishes to risk it all on one throw of the dice can get no help from probability theory in deciding whether to bet on a Seven or on Snake Eyes.

6   ABSOLUTE PROBABILITY AND PHYSICAL CHANCE

Reichenbach introduces the term 'absolute probability' and specifies its meaning as follows:[51]

> The probability $P(B)[= df\, P(A \lor A, B)]$ may be called an absolute probability, in contradistinction to the relative probabilities so far considered. An absolute probability can be regarded as a relative probability the reference class of which is the universal class.

No analogous idea exists in von Mises' theory. For him, when we speak of *the* probability of *B*, we are implicitly referring to an appropriate and limited reference class (collective) rather than explicitly referring to the universal class. We have already discussed his argument that there is no such thing as the probability that Mr Smith will die in the new year because Mr Smith belongs to many different collectives, each of which may have a different death-frequency.[52] It is likewise true that there is no such thing as an *intrinsically* improbable single event. Consider, for example, what he has to say about an 'unusual' lottery number winning the prize:[53]

> The event that the first prize will fall to ticket No. 400,000 has, in itself, no 'probability' at all. A collective has to be defined before the word probability acquires definite meaning.

Here our surprise at this 'improbable event' is due to the fact that one ticket which exhibits a very rare property in one collective (a winning ticket in a collective of lottery tickets) *also* exhibits a very rare property in another collective (a number ending in 5 zeros in a collective of numbers). To put von Mises's point another way, getting a 'round' number and getting a winning lottery ticket are rare but independent events. One would indeed be foolish to bet that the number he will draw will be both 'round' *and* a 'winner,' since the odds would be the *product* of the two small probabilities. But if one *has* a 'round' number, one would be equally foolish to throw it away in the belief that such numbers never win, since, of course, they win as frequently as any other number in a fair lottery. When someone holding such a ticket does indeed win we think of it as a rare event. But it is not the *event* that is rare, but its properties. The ticket with which we are concerned has two rare properties: being a winning ticket and having a round number. But it *also* has properties which are *not* rare, such as rectangularity, redness, having 4 for the first number, being sold in New York, etc. If we were concerned with *these* properties, considered in their proper collectives, the event would not be rare at all. Since the rarity of an event depends on the rarity of its properties, and the rarity of a property depends upon the collective in which it is being considered, the only use of 'the probability of *X*' is to refer to *the* probability of a specified property in a specified collective. There is no such thing as *the* probability of an event.

Keynes did specify a sense for the phrase 'the absolute probability of *p*,' remember, when he said:[54]

If *h* specifies the Universe of Reference, i.e., if its group compre-hends the whole of our knowledge, *p/h* is called the *absolute probability of p*, or (for short) the probability of *p*; and if *p/h* = 1 and *h* specifies any real [i.e., known to be true] group, *p* is said to be *absolutely certain* or (for short) *certain*.

This definition of Keynes's does capture one concept to which we may be referring when we speak of 'the absolute probability of *P*' or simply '*the* probability of *P*.' This concept is the probability of *P* relative to everything we know, or in light of all relevant evidence.

There is another sense, though, especially common to scientists and metaphysicians, which differs from this and is not captured by Keynes. This is the notion of the *real* probability of *P*, or the probability determined by what is true of the universe and not by what is known. It is said to measure the actual chance, or likelihood, that an event will occur.

It would seem that Reichenbach's definition is addressed to this latter sense, because his reference class is not our body of knowledge but the actual universe – everything which is either *A* or not-*A*. Indeed, the very definition of probability used by Reichenbach requires that the absolute probability be an objective fact, indepen-dent of our knowledge and dependent on the nature of the real world. But what do we have when we have this absolute probability? Well, if we substitute into Reichenbach's definition, we find that we have the number of things which satisfy *B* divided by the number of things which satisfy *A* or not-*A* (or rather, the limit of this ratio as the population expands to infinity). This is useful information if we wish to know how many *B*'s there are (or ever will be) in the universe, and how many other things; but does it really tell us what the probability of *B*-dom is? If we ask for the absolute probability of the existence of telepathic communication devices, are we satisfied when told that 0 is the ratio of the number of such devices (or any finite class of things) to the infinite (?) number of things in the universe? Surely *that* doesn't answer our question! And if we ask for the absolute probability that we shall encounter the department chairman today, and that request is answered by the ratio of such encounters to the number of events in the universe (again zero), we are at a loss for something to say. The problem seems to be specifying what counts as a relevant thing but not as a *B* – that is, the specification of some reference class to give meaning to our query.

175

But when we *do* specify such a class, we have left absolute probability and are again dealing with relative probability. Perhaps there is no way to make sense of 'absolute probability' in objective terms! But let us look to see how Reichenbach uses his concept, how he deals with absolute probabilities.

He doesn't.

After giving his definition, he notes that if a sequence is 'compact' in some class A (i.e., every member of the sequence is a member of A) then A is the equivalent of the universal class $(A \lor \bar{A})$ for that sequence. Thus when we use A as the reference class for determining the probability of B, $[P(A, B)]$ we may omit reference to A and speak directly of the (absolute) probability of $B[P(B)]$. In other words, the only use Reichenbach makes of his definition of absolute probability is to authorize us not to state the reference class in a limited universe of discourse. What it would mean to do this in an unlimited universe is never explored.

This notion of suppressing the evidence or restrictions in a limited universe of discourse ('The probability of drawing an Ace is 1/13') has been common to probability theory since its inception; even von Mises allows *this*. We must conclude that despite Reichenbach's talk of absolute probabilities he has not introduced a potentially fruitful new concept (whereas Keynes's concept *is* meaningful, for example).

It would seem fairer to the spirit of Reichenbach's theory to ignore his brief remarks on 'absolute probability' and instead to concentrate on his continued insistence that all probabilities are relative since they essentially involve the frequency of some property in some reference class.[55] It is both true and important that Reichenbach denies that any 'absolute probabilities' can be given which are not relative to some reference class.

But the rejection of absolute probabilities *in this sense* emphatically does not mean that Reichenbach thinks probabilities are in any way relative *to human knowledge*. Quite the contrary – he completely agrees with von Mises's thesis that probabilities are objective properties of the physical world and like mass and length obtain completely independently of human language or knowledge.[56]

> If the reference class is stated, the probability of an attribute is objectively determined, though we may be mistaken in the numerical value we assume for it on the basis of inductions.

The probability of death for men 21 years old concerns a
frequency that holds for events of nature and has nothing to do
with our knowledge about them, nor is it changed by the fact
that the death probability is higher in the narrower class of
tuberculous men of the same age.

The interesting thing about such 'objective properties' is that it
follows that every probability is *an unchanging physical fact about
the universe*. If there *is* a limiting value to the relative frequency of
occurrence of red giants in the stellar population, there can be only
one such value, determined by the entire history of the universe.
This value has nothing to do with human beings, their language, or
their knowledge. In this sense we may wish to say that RF
probabilities are more nearly absolute than AP probabilities.

It is unlikely that RF theorists would be happy with this talk of
absolute, unchanging RF probabilities. One of the virtues which
they profess for their theory is that it can learn from experience and
can change, for example, the probability that a die will yield a Five,
while Classical theories especially cannot take note of experience.
Here I think the RF theorists are guilty of confusing our *judgment*,
belief, or *estimate* of the probability with the probability itself. For
example, at one place[57] von Mises talks about how we establish a
figure for the probability of death among 40-year-olds and then
about how the figure might change:

no figure of this kind, however exact at the moment of its
determination, can remain valid forever. The same is true for all
physical data. The scientists determine the acceleration due to
gravity at a certain place on the surface of the earth, and
continue to use this value until a new determination happens to
reveal a change in it; local differences are treated in the same
way.

But there are two reasons why scientists might be led to change
their figures:

1 More precise measurement might give a closer approximation
to the true value of G.
2 Physical changes in the earth might cause a change in the
true value of G.

Of these, only the first can apply to a probability, for if a limit
exists to any empirical frequency, it will remain always the same.

No amount of divergence in the observed value up to any finite point can alter the fact that the series will eventually 'correct' itself and converge on the final value. We must be clear, as von Mises is not in this passage, that the only values that can *change* are our estimates or opinions or beliefs about what the true probabilities are – the probabilities themselves as von Mises defines them are eternally immutable if they exist. The reason for this fundamental difference is that probabilities are not like ordinary physical properties *even on von Mises' theory*, because (1) they refer to classes of individuals, properties, or events, and (2) these classes are defined as extending into the infinite future. Since all future events of type A are *included* in the limit process, none of those events can *change* the limit. Therefore no future event or series of events can change a present probability.

Now we take up the question of physical chance. There is nothing in the RF theory itself to settle this question one way or the other. Von Mises clearly believes in the existence of physical chance in the sense that he thinks a Laplacean determinism is unobtainable. It is less obvious whether he thinks this is due to the practical difficulties of such predictions or whether he thinks the world is so undetermined that it is not even in principle predictable. That question would perhaps be answered if we knew whether or not there are empirical sequences which meet his requirement of randomness. Such sequences might seem to fulfill our ordinary criteria for chance or random processes.

It is clear that von Mises thinks there are such sequences, and that they represent both the subject-matter and the justification of his theory of probability. Games of chance and actuarial regularities are examples he cites again and again.

It is less clear, though, whether he thinks such sequences are *in principle* unpredictable, because they are 'truly' random processes, or whether, like Poincaré, he thinks only that they are unpredictable, in practice because of the minute observations and vast calculations which would be required to achieve reliable solutions. The earlier reference to 'an easily recognizable law' might tend to support the latter interpretation. There is also rather extensive discussion of and support for Poincaré's notion that the 'chance' in 'chance mechanisms' arises from small causes which produce large effects,[58] which is often used to explain why prediction is impossible in a deterministic world.

Other passages, however, affirm the alternative interpretation. On page 26 he introduces a 'Principle of the Impossibility of a Gambling System' which asserts that there is no 'selection principle' which can pick out in advance the successes in a random sequence of trials. This Principle is explicitly analogous to the Law of the Conservation of Energy, and its forbidding of a gambling system is said to be similar to the impossibility of a perpetual motion machine. The unpredictability thus seems to be theoretically and not just practically ineliminable.

This interpretation receives further and, I think, decisive support from the final part of the book, where von Mises discusses statistical physics and the rise of quantum mechanics.[59] He comes to the conclusion that the inexact, statistical nature of *all* measurement and the restrictions of the Heisenberg Uncertainty Relations together show that the precise data required by classical determinism are in principle unobtainable.[60] As a result, he urges that we renounce the 'prejudice' of determinism and either modify or abandon the notion of causality.

From all of this we may conclude that von Mises believes in the existence of physical chance in the ordinary sense we have been using even if we have doubts about the theoretical adequacy of the grounds of his belief.

Two final points about this matter. First, the existence of physical chance is taken to be contingent and the Principle which guarantees it is merely an empirical generalization, very far-reaching and extremely well supported by experience but devoid of self-evidence and insusceptible of deductive proof. In this, as in other matters, von Mises is a properly anti-metaphysical Logical Positivist. There are no necessary truths about his physical universe.

Second, the requirement of randomness as formulated by von Mises has been found unsatisfactory by most subsequent theorists.[61] For one thing, it rules out *every* empirical sequence if applied strictly enough, because there *is always* a method of selecting the successes in an empirical sequence, namely, any method or formula which in fact happens to pick out the relevant place numbers, even if it is only a list of those numbers. It is true that such methods are generally unknown to us and at any given time cannot be expected to have continued success in the future. But this argument again seems to indicate that the impossibility of prediction is practical only, rather than theoretical.

179

I think it would be a misapprehension to take this technical criticism as having any metaphysical significance for the following reasons. Von Mises has said that a series is random if there exists no selection system which would pick out a subset with a different limit of the RF. His critics have argued that such a method always exists, even if it usually is only an enumeration of the subset, rather than a constructive generation of it. But to say that there is a variant subset is to say only that the collective is not homogeneous. And to say that there is a true description of that subset is no more metaphysically (or pragmatically) significant than the corresponding assertion that there 'is' a true description of the future course of the universe. This latter assertion is clearly compatible with indeterminism (if one thinks the true description is not yet distinguished or selected from the class of all possible descriptions), or theoretical determinism (if one thinks the true description can be ascertained, at least in part and theoretically *in toto*, by investigation in the present and near future). Analogously, the fact that we know beforehand that we will be able to describe the subset after its occurrence (and therefore that such a description 'exists') has no bearing on the question of whether the sequence is random or deterministic.

Reichenbach's theory is also compatible with the Laplacean notion that probabilities (here, frequency summaries) are useful ways of overcoming our inability to grasp successfully the true determinism in the world. This latter alternative is suggested by Reichenbach's analysis of randomness, for example, which concludes that von Mises was wrong in taking randomness as a necessary condition for the application of probability theory. Instead, he specifically says that the impredictability of a roulette wheel springs from our inability adequately to observe and extrapolate the physical data. 'But this is true only in view of the limited abilities of human observers,' he says, 'Laplace's superman' would not be subject to such 'psychological randomness.'

From this we may conclude that the world is predictable rather than random. However, this conclusion validly applies only to Reichenbach's view of the *macroscopic* world. In the world of subatomic physics, Reichenbach thinks that the randomness of events is a genuine property of the physical world. Because of this, probability has replaced causality in modern physics to the point that it is now essential to the correct description of realty. The apparent discrepancy between microscopic randomness and macroscopic determinism is dissolved by the Law of the Compensation of the Dispersion. This general statistical law is based on the principle that 'for non-vanishing dispersions, the linear dispersion of the sum is smaller than the sum of the linear dispersions.' As a result of this mathematical feature, a large deviation by one individual in one direction is often balanced or compensated by an equal deviation of another individual in the opposite direction. This is why businesses can rely on 'average sales' in their calculations. But in the microscopic world the number of individuals is *so large* in comparison to the possible deviations that such ensembles exhibit regular behavior with virtual certainty. This compensation guarantees predictability despite the general randomness espoused by quantum mechanics. Thus there is genuine randomness, but *most* probability predictions are, like Laplace's, substitutes for unattainable but intrinsically possible deterministic predictions.

7 METAPHYSICAL STATUS OF *P*

In von Mises's theory an elementary probability statement is said to be an ordinary scientific statement of a physical fact, just like statements of specific weights, masses, etc. Our *knowledge* of such statements is obtained through normal, inductive science;[62] their *truth* depends upon the contingent nature of the universe. If we let *P* be 'The probability that a star is a red giant is 0.02,' then we *learn* the truth of *P* through astronomical observations and its truth means that the (pseudo-) infinite sequence of stars contains red giants in a proportion that approaches 0.02 as a limit.[63]

Many people object to this theory on the grounds that limit statements are not like 'ordinary' scientific statements at all and their (meta-) physical interpretation presents difficulties not found in statements like *S*: The specific weight of this wood is 0.08. But von Mises thinks that *P* is like *S*, not because there is an obvious and simple physical interpretation of *P*, but because *even statements like S are limit statements.* He says:[64]

> I have mentioned on many occasions that all the results of our calculations [of probability] lead to statements which apply only to infinite sequences.... It might thus appear that our theory could never be tested experimentally.

This difficulty, however, is exactly the same as that which occurs in all applications of science.

As an example, suppose we wished to verify experimentally the statement *S*, above. We would naturally measure out a volume of the wood and weigh it. But von Mises says this is not sufficient because

> the weight of only a finite volume of the substance can be determined in this way. The value of the specific weight, i.e., the limit of the ratio weight/volume for an infinitely small volume, remains uncertain just as the value of a probability derived from the relative frequency in a finite sequence of observations remains uncertain. One might even go so far as to say that specific weight does not exist at all, because the atomic theory of matter makes impossible the transition to the limit of an infinitely small homogeneous volume.[65]

I am not too happy with this argument because when I took College Physics, specific weight was defined as the weight *per unit volume*, with no limiting process involved or implied. My (possibly outdated) physics texts and dictionaries still define it that way. But we need not pursue this question because von Mises goes on to consider the meaning and possible verification of 'the indication of the weight of a certain finite volume of matter,' which, for unit volume, is (roughly) equivalent to my definition of specific weight *sans* limit. Even this statement, he says, is not as transparently meaningful as we might think.

> as soon as we begin to think about a really exact test of such a statement, we run into a number of conditions which cannot even be formulated in an exact way. For instance, the weighing has to be carried out under a known air pressure, and this notion of air pressure is in turn founded on the concept of a limit.[66]

He later goes on to argue that *all* physical measurements must 'form collectives' because macroscopic objects are only statistical collections of particles while microscopic objects are governed by Heisenberg's Uncertainty Relations. Therefore exact measurements are always impossible and only limiting processes are appropriate in physics.

The purpose of this entire argument (which is of the *ad hominem* subspecies known as *tu quoque* or as some call it, 'so's your old lady') is to show that *all* physical statements require the notions of limit and infinite series in their interpretations, and therefore RF probabilities are *no worse off* than other statements in this regard. This line of attack obviously turns heavily on the verificationist (or operationalist) tenet that the *meaning* of a physical statement (or quantity) depends on how it is verified (or measured) in practice. Those who accept this principle will presumably be convinced, while those who reject it may continue to believe that there is a significant difference in meaning between statements which explicitly refer to the limiting value of a relative frequency in an infinite series and those which do not.

Whatever the difficulties of interpretation and verification, it is clear that *P* is intended to be an empirical statement which says something about the physical universe in exactly the same way as other statements of scientific fact. Because of this, von Mises repeatedly asserts that his is an *empirical* theory which is properly based on observation. This essential empiricism can be seen in one last statement of his concerning probability in dice games:[67]

> The probability of a 6 is a physical property of a given die and is a property analogous to its mass, specific heat, or electrical resistance.

In discussing Reichenbach's views on this topic, it is essential to keep in mind the distinction between abstract, theoretical statements and statements of the applied or interpreted system. After all, Reichenbach probably did more to elucidate and emphasize this distinction than any other modern philosopher (with Carl Hempel and Ernest Nagel as possible exceptions). Therefore we shall speak here of two kinds of statements, rather than the familiar, all-inclusive '*P*' for probability statements. We shall use 'PT' to refer to statements of Probability Theory, that is, those axioms and their consequences which constitute Reichenbach's version of the probability calculus. The term 'PA' will stand for 'Probability Application' and will represent all those statements which assert a real probability value. This distinction should be familiar to everyone who has read any philosophy of science and readily understood by anyone who grasps the difference between pure and applied mathematics.

As might be expected, the two types of statements have different metaphysical status. The first group, PT, includes a body of invented axioms designed to possess fecundity in the generation of deductive consequences. It would be idle to pretend that such axioms are not suggested by empirical considerations or that they are not intended to produce a system of one kind rather than another. But the important thing to recognize about the system of statements PT is that it is a *formal system* in the now familiar logico-mathematical sense of being a deductively closed system resulting from a specified set of axioms, definitions, and undefined terms. As such, it says nothing about the physical world. Every statement PT is a *tautology* in the sense that its truth is guaranteed by the laws of logic and the probability axioms. Reichenbach rejected von Mises's claim that the probability calculus arises by induction from observation as do the laws of natural science – rather, he saw it as resembling a system of geometry, whose axioms and consequences may or may not have a reasonable interpretation in the real world.[68]

But Reichenbach's axioms do not stand in splendid isolation, governing an independent realm. Instead, he shows that they follow logically from the mathematical theory of infinite sequences. That is to say, if one stipulates that probabilities are to be interpreted as limits of relative frequencies, then the axioms and theorems of the probability calculus become ordinary mathematical truths. It is an important consequence of Reichenbach's theory, therefore, that the system PT is not just whimsical but has the same assertability as higher mathematics. Still, we must remember that Reichenbach has himself insisted that the truths of mathematics are analytic and therefore metaphysically empty. Probability statements of the PT type say nothing about the existence or properties of any real object and therefore have no metaphysical significance.

Quite the contrary is the case for the applied probability statements, PA.

As should be evident by now, Reichenbach thinks that the formal system PT is important because, when applied to the world, it helps us derive new synthetic truths from old ones. It functions exactly like the formal systems of arithmetic and geometry, as can be seen by the following examples:

A1 I am putting 2 apples in the bag.
A2 There were already 2 apples in the bag.
A3 $2 + 2 = 4$.

A4 There are now 4 apples in the bag.

G1 This piece of land is a right triangle.
G2 This side is 3 km long and that one is 4 km long.
G3 Pythagorean Theorem: $a^2 + b^2 = c^2$.
G4 The other side must be 5 km long.

P1 The probability of drawing an Ace is 1/13.
P2 The probability of drawing a Ten is 1/13.
P3 For independent events, $P(A \vee B) = P(A) + P(B)$.
P4 The probability of drawing an Ace or a Ten is 2/13.

Under the standard interpretation, A3 is an analytic truth of arithmetic and G3 an analytic truth of Euclidean geometry. In Reichenbach's system, P3 is an analytic truth of the probability calculus. None of these says anything about reality.

Statements A1 and A2 are statements of applied arithmetic. They are synthetic assertions that certain real things have certain numerical properties. For Reichenbach, P1 and P2 have the same synthetic status. They assert that the empirical sequence of card draws includes draws of Aces and Tens with relative frequencies which will each approach 1/13 as a limit if the sequence is indefinitely extended. Such statements may be either true or false, and their truth or falsity is determined by facts in the real world rather than facts about language and logic. They are therefore synthetic propositions. (The reader may reasonably have doubts about the synthetic status of statements about card games. If so, one could replace the example with one about death frequencies which are clearly contingent empirical facts.)

Finally, statements A4, G4, and P4 are synthetic statements which are asserted as the consequences of their respective premises. It is important that they be synthetic, because the entire pragmatic function, the practical purpose, of arithmetic, geometry and probability theory is to enable us to arrive at previously unrealized truths about the world.

Now synthetic statements *do* have metaphysical import: they claim to be true descriptions of some part of reality. In particular, P4 claims that the ratio of successful draws to this straight will eventually average out to (approximately) 2/13.[69] By knowing this, we are better equipped to assess our prospects and apportion our assets. Applied probability (PA) statements help us to deal with the future by

185

describing the general proportions which will obtain in future events. There remain difficult questions about the application of this general knowledge to particular circumstances, but few will deny that knowledge of the broad outlines and proportions of the future is better than no knowledge at all. PA statements are thus important because they claim to describe physical reality.

At the risk of confusing the issue a bit, I would like to remind the reader that statements about the probability of single events should *not* be interpreted as asserting some claim about the nature or description of such events. They are, after all, not genuine statements at all, but pseudo-statements. When I claim that the probability of my filling *this* straight is 2/13, I do not mean to imply that the future event is somehow metaphysically indeterminate or undetermined or possessed of any specific property. The *real* content of my assertion (according to Reichenbach) is that if I posit or act upon the belief that I will fill the straight, and if I repeatedly do so each time such an occasion arises, then the ratio of my successes in the long run will be 2/13. I have no way of knowing if *this* draw will be one of those successes or not, and my statement should therefore not be construed as applying to *just this draw*, but rather as an implicit description of the series of all such draws. When reinterpreted in this manner, statements about the probability of single events can be seen to be disguised statements about proportional distributions in the real world and therefore have the same metaphysical import as any PA statement. If interpreted *literally*, however, they are strictly nonsensical and make no metaphysical claim.

Finally, we should consider the metaphysical significance of Reichenbach's system as a whole and its application to reality. The reduction of the probability axioms to truths about the mathematical properties of infinite series has shown, according to Reichenbach, that the application of his probability theory to the world is justified if and only if the Rule of Induction is. We will return to this claim in the section on the Rationality of Probability Behavior. For now we are interested only in what is *being* claimed, not whether the claim is justified.

The fundamental assumption, principle, foundation, or justification of Reichenbach's system is the proposition that the empirical world is characterized by systematic regularities which disclose themselves in empirical sequences. That is, at least some sequences of events (and therefore observations) in the real world do tend to stabilize in certain proportions. This is obviously a metaphysical claim of the first magnitude. It is similar to the traditional principles that Nature is Uniform and The Future Will Resemble the Past. The important differences are that (1) Reichenbach describes the uniformities as being equivalent to the existence of limits for empirical sequences, and (2) the uniformity is asserted hypothetically only.

The first of these points allows for the metaphysical indeterminism of particular events while asserting that there can still be detectable regularities in nature. The second says that we can never know or justifiably assert that reality is uniform in this sense, but Reichenbach claims, *if* we have *any* hope for success in dealing with the world, we can attain that success by *assuming* regularity and acting accordingly. If there is no regularity, we are foredoomed to failure anyway, but if there is regularity (and the observation of *portions* of such sequences suggests that there is) then the RF probability theory will eventually disclose that regularity and lead to success in practice.

8   THE EPISTEMOLOGICAL STATUS OF P

Von Mises has repeatedly stated that his is an empirical theory, and denigrated the apriorists for their rationalistic approach of establishing laws of probability by logical considerations rather than through observation and experiment. He gives this picture of how a probability theory should function:[70]

> Like all the other natural sciences, the theory of probability
> starts from observations, orders them, classifies them, derives
> from them certain basic concepts and laws and, finally, by
> means of the usual and universally applicable logic, draws
> conclusions which can be tested by comparison with experimen-
> tal results. In other words, in our view the theory of probability
> is a normal science, distinguished by a special subject and not
> by a special method of reasoning.

That certainly represents a traditional description of the nature of an empirical science. Unfortunately, it does not jibe with von Mises's detailed description of how *his* theory is to be tied to experience.

In the first place, remember that he argues that the establishment of initial probabilities is *not* part of the task of a probability theory (while I have argued that it is the *most important* such task), that only the manipulation of such data is proper to probability theory. In the second place, the laws of his theory are *not* derived from observation, but deduced from his premises. Thus neither the data nor the laws of the theory are necessarily empirical.

To support the first contention, we will briefly review what von Mises had to say about initial data. For example:[71]

> In a problem of probability calculus, the data as well as the results are probabilities.

This is an unexceptional remark which points up the purely mathematical character of the probability calculus – it accounts for the widespread agreement among divergent theories on the question of what can be done once initial probabilities are established. But von Mises goes much further than the others – he applies these restrictions to the entire *theory* of probability.

> A great number of popular and more or less serious objections to the theory of probability disappear at once when we recognize that the exclusive purpose of this theory is to determine, from the given probabilities in a number of initial collectives, the probabilities in a new collective derived from the initial ones.[72]

> The task of the theory of probability is to derive new collectives and their distributions from given distributions in one or more initial collectives. The special case of a uniform distribution of probabilities in the original collective ('equally probable' cases) plays no exceptional role in our theory.[73]

The task of calculating the original probabilities is assigned to statistics. Von Mises draws a specific analogy between the statistician's duty to furnish probabilities for the probability theorist to operate on and the surveyor's responsibility to provide initial measurements for the geometer to use in his calculations. In each case the calculations will proceed according to fixed rules and arrive at a 'correct' answer regardless of the degree of reliability of the original data.

I think this demonstrates that von Mises's theory does not include any workable procedure for calculating initial probabilities, nor even any effective rule that they must be empirically determined (statisticians, too, can be apriorists; indeed, some would argue that when they ascribe a limit to an infinite empirical series they *must* be apriorists, as the data alone are always insufficient for the conclusion).

So it is not the case that von Mises's version of the RF theory is empirical in the sense that its data (initial probabilities) must be based on experience. It remains only to substantiate our claim that the laws of the theory (the probability calculus) are also not 'based on observation.' In this matter, we are in complete agreement with Carnap:[74]

> Mises has repeatedly stated ... that his theory of probability is empirical, is a branch of the natural sciences like physics. However, his theorems, although referring to mass phenomena, are quite obviously purely analytic; the proofs of these theorems (in distinction to examples of application) make use only of logicomathematical methods, in addition to his definition of 'probability', and not of any observational results concerning mass phenomena. Therefore his theory belongs to pure mathematics, not to physics. This point has been discussed in detail and completely clarified by F. Waismann ...

Indeed the only place in which I have found von Mises referring to observational and experimental evidence in support of his theory is in his attempt to show that his assumption of the existence of limits is a reasonable one. As the 'experimental basis' and justification for this theory, von Mises cites the enormous documentation of Monte Carlo and other such gambling houses, which shows that each game always tends to a limit, together with the similar results of state lotteries. He concludes:[75]

> We thus see that the hypothesis of the existence of limiting values of the relative frequencies is well corroborated by a large mass of experience with actual games of chance. Only processes to which this hypothesis applies form the subject of our subsequent discussion.

The question of which processes *do* fulfill the hypothesis is one that cannot generally be settled on either an empirical or a theoretical basis. It is a problem in the application of probability theory. For particular processes, there are three possible solutions. (1) The process is governed by an obvious causal or mathematical regularity which makes it predictable. This case is excluded from von Mises's theory by the randomness condition. (2) The process is *defined* as purely random or possessing a definite probability which makes it statistically but not individually predictable. This case is treated identically by von Mises and the apriorists, since only the probability calculus is involved. (3) The process is not *known* to have a governing probability or causal mechanism but has thus far exhibited statistical regularity. Von Mises says we can deal with this case by *assuming* it has a limit which is ascertainable by the statistical results.

If, in the later course of experience, we find the process repeatedly and significantly violating the prediction of the theory, *it does not falsify the theory.* Instead, says von Mises, 'it would indicate that this sequence of observation does not satisfy the conditions of a collective.'[76] (Isn't this reminiscent of Keynes's rejection of Wolf's experiments on the grounds that his dice must not have been 'fair'?)

If any of this involves the empirical justification of probability laws, it is not glaringly obvious. The laws are deduced from von Mises' definitions and axioms.[77] The application of the laws is doubtful in particular cases, and is generally justified only by the perceived fact that many processes have *thus far* exhibited marked statistical regularity. The assumption that they and similiar processes will continue to exhibit such regularity in the future is not logically superior to the use of 'The future will resemble the past' and other uniformity principles to justify induction. In particular, this justification-by-regularity applies just as well to successful a priori theories and inductive logics as it does to von Mises's RF theory.

If the initial data of von Mises's probability theory are not necessarily empirical but logico-mathematical, does there remain *any* sense in which it is proper to describe this theory as epistemologically empirical?

I think that we can answer this question affirmatively, as von Mises clearly wishes, if we again judiciously ignore what he said about the theoretical irrelevance of initial probabilities. If we stress instead his argument that initial probabilities cannot be based on logical considerations (and especially not on the equal distribution of our ignorance) but must result from actual observations over a period of time, it is clear that such statements are indeed empirical. If we allow, as is usual, that the property of being empirical is hereditary through both inductive and deductive inference, then all subsequent probability statements which are based upon the initial ones must also be empirical. But this includes virtually all the statements made by probability theory except the axioms and theorems and their analytic consequences. A large part of von Mises' theory, therefore, can properly be called 'empirical.'

There is even, I think, a sense in which the axioms and theorems themselves might be described as 'empirical.' This is the sense in which a *theory* may be described as empirical if it is *intended* to be modelled in experience, or, even better, if at least one empirical model exists which successfully approximates the behavior of the theoretical entities. I take it that the proper use of von Mises's examples of games of chance, actuarial statistics, and physical processes is to show that his theory is empirical in this sense. But now we must be careful, because the Classical and AP theories are nearly as empirical – in this attenuated sense – as RF theories are, because they too have successful models in experience (though perhaps not *as many* models). In this sense any mathematical formalism which successfully deals with concrete reality can be described as empirical – even deductive logic, perhaps. It would be better, I think, if we base von Mises's claim to empiricism not on this characteristic of his theory (it is really a property of the probability calculus *in abstracto*, anyway) but upon the fact that initial probabilities are based upon empirical, rather than logical or intuitive, considerations.

In discussing Reichenbach's views, we shall continue to emphasize the difference between those statements of probability theory (PT) such as the laws and the calculus of probability, and those statements of probability applications (PA) which assert the value of a probability in the real world. As one might expect from the metaphysical differences discussed above, the two types of statements also (consequently) have significantly different epistemological status. To put it briefly in philosophical jargon, PT statements are analytic a priori while PA statements are synthetic a posteriori.

The PT statements are analytic because they are deductive consequences of the mathematical laws of infinite sequences, which in turn (we assume with Reichenbach) are derivable from ordinary deductive logic. They are therefore logical truths – hence analytic.

Most philosophers would agree that analytic truths can be known a priori (i.e., that their warrant or verification does not depend on the nature of experience or reality) since they deal only in logico-linguistic relations and say nothing about the world. In Reichenbach's words:[78]

[the] axioms of the calculus of probability follow tautologically from the frequency interpretation. Therefore, if probability is regarded as meaning the limit of a relative frequency, the validity of probability laws is guaranteed by deductive logic. This result is of major importance for the epistemological critique of probability statements.... Now the assertability of theorems of deductive logic is based on the fact that logical formulas are empty, that they do not anticipate properties of the physical world. In the same sense, therefore, theorems of probability must be regarded as empty when applied to relative frequencies. The theorems constitute mere transformations of probability expressions into others, without any addition as to content.... That, none the less, the conclusions of probability transformations can be new, in the psychological sense, is obvious to the logician who is familiar with the twofold nature of tautologies as logically empty but full of psychological content.

Now Reichenbach thinks that the truths of logic and mathematics are 'certain'[79] and 'necessarily true.'[80] Therefore we conclude that the epistemological status of probability statements of the PT type is that they are among our greatest certainties, possessing the full and compelling epistemological warrant of deductive logic and mathematics.

Unfortunately, things are not quite so satisfactory for PA statements in Reichenbach's theory. You will recall that these statements assert the existence of some probability values in the world. They are therefore clearly synthetic, since their predicates are not 'contained in' their subjects but 'add to' them informatively. ('The probability of death in the 30th year' does not mean the same as '0.005' or necessarily equal it. The statement that it *is* 0.005 is therefore a synthetic statement.)

I think it is fair to say that most philosophers in the Anglo-American tradition reject the possibility of synthetic a priori knowledge. But even those who insist on the existence of *some* synthetic a priori knowledge – even Kant – will not claim that our knowledge of death frequencies is a candidate for that honor. So we may conclude that our knowledge of Reichenbachian probabilities must be a posteriori, as well as (or because) synthetic. This is, of course, the same broad classification which includes all our scientific knowledge, our common-sense beliefs about the world around us, and, in general, any of that great mass of knowledge that is 'based on experience' rather than logically derived.

This is not bad company for PA to be in. After all, empiricists have long been claiming that all *important* knowledge is in this class, and that most of it is highly reliable (even certain, some say, for some parts). But so far we have not come to the real difficulty – all synthetic a posteriori statements are not created epistemologically equal, and Reichenbach's PA statements suffer from a particular built-in problem common to von Mises' and other RF systems: infinity.

We are asked to believe that the meaning of the statement 'The probability that a star is a red giant is 0.05' is that the relative frequency of red giants among observed stars approaches 0.05 as a limit when the number of observed stars approaches infinity. Suppose we do accept it as an analysis of meaning. On what grounds would we ever accept it *as a matter of fact*. One doesn't just 'look to see' what properties an infinite empirical sequence has. Nor can one analyze the properties by a recursive enumeration of the series; while this is possible for mathematical series, it is naturally impossible for empirical series. This difficulty is so great that it counts as one of the major criticisms of RF theories in general, and will be discussed later. Here we will merely present Reichenbach's position about the epistemological status of PA.

Reichenbach accepts as true the proposition that all PA statements are synthetic and hence not provable by means of the calculus alone,[81] that the limit of an extensionally (empirically) given sequence is not verifiable,[82] and that in any given case we may be mistaken in our belief about the numerical value of a probability.[83] But none of this discourages him, because he thinks it just shows that probability beliefs are like all other scientific beliefs in being less than certain but inductively well-founded. Indeed, the case can be put even more strongly: the assumption that an observed rf will continue in the future is not just inductively well-founded, it *is* induction.

> we cannot know, strictly speaking, towards what limit such a sequence will proceed. We assume, however, that the observed frequency will persist, within certain limits of exactness, for the infinite rest of the sequence. This inference, which is called

inductive inference, leads to very difficult logical problems; and it will be one of the most important problems of this investigation to find a satisfactory explanation of the inference.[84]

The upshot of Reichenbach's explanation is that PA statements are obtained by using the Rule of Induction. It follows, then, that they are epistemologically well-founded if, and only if, that Rule is. We address this problem in the following section.

9 RATIONALITY OF PROBABILITY BEHAVIOR

Von Mises has little to say about the rationality of probability behavior. This is apparently because he thinks of himself in the role of scientist rather than philosopher; his goal is a true description of the world; what we do with it is our business.

Indeed, there is some doubt about how one *could* act on von Mises' description of the world. He warns us against applying the theory to any particular case, and I have argued above that there is no clearly satisfactory way in which it can be applied to multiple but finite groups.

Despite these difficulties of interpretation, it seems clear that von Mises' attitude is that it is rational to act on the best possible description of the world and that his theory is a great improvement over earlier versions and gives very good predictions indeed. Therefore it is rational to act on his RF theory of probability.

Reichenbach, on the other hand, has devoted a good deal of thought to this question and has in fact developed one of the most important arguments in favor of the rationality of probability behavior, to which we now turn.

While Reichenbach describes many complex statistical and probabilistic models for use in a state of 'advanced knowledge,' we need not worry about these complications. In the end, all such elaborations are based upon the simple method of induction by enumeration, and are justified if, and only if, this method is justified in a state of 'primitive knowledge.'[85]

Induction by simple enumeration is said to proceed according to the

RULE OF INDUCTION. If an initial section of n elements of a sequence x_i is given, resulting in the frequency f^n, and if, *furthermore, nothing is known about the probability of the second level for the occurrence of a certain limit p, we posit that the frequency $f^i(i > n)$ will approach a limit p within $f^n \pm \delta$ when the sequence is continued.*[86]

This is, of course, the Straight Rule of Induction, advising us always to assume that an observed series will continue to exhibit the same relative frequency. This rule is so fundamental to Reichenbach's theory that he describes it as the only non-analytic assumption required for the validity of probability theory. He contends that the rest of the RF interpretation of probability follows logically from the assumption of this rule and the definition of probability in terms of relative frequency. We can therefore say that, for Reichenbach, *probability behavior is rational if and only if the Rule of Induction is justified.*

Now clearly Reichenbach doesn't think that the Rule of Induction is justified in the sense that each application of the rule must lead to a correct answer. It is perfectly plain that application to the same series at different times will yield different values for the projected frequency (posits) unless the series is absolutely homogeneous. But '[it] is not the use of an individual [posit] but the progressive use of the total set that can be shown to be advisable; such progressive use will lead to a prediction of the frequency if there is a limit to a frequency.'[87]

That is to say, if we toss a coin once, and get Heads, we posit that the frequency of Heads is 1. After a second toss, we posit 1/2. Then 2/3, 3/4, 3/5, 3/6, 4/7, 4/8, etc. If this series has a limit (say 1/2) we must eventually reach an event number, N, beyond which the posited value differs from the true value by no more than an arbitrarily small amount.[88]

For Reichenbach, the existence of limits to empirical sequences is a necessary condition for the predictability of experience. (It is also a sufficient condition if one employs the Rule of Induction.) Thus, success in predicting the future is possible only if at least some empirical series have limits. But if they *do* have limits, then it is a mathematical certainty that the Rule of Induction must eventually discover those limits and thereby succeed in predicting the future. We therefore have reached Reichenbach's famous Pragmatic Vindication of Induction: *Use of the Rule of Induction will lead to success in predicting the future, if success is attainable at all.*

This is called a 'pragmatic vindication' rather than a 'justification' of induction because it does not claim that induction *will* lead to success. Indeed, Reichenbach asserts that we *cannot* prove that the success of induction is necessary – or even probable.[89] All we can show is that it will succeed if anything will. But this is sufficient (in the absence of any clearly *better* method) to justify the use of induction as rational. We are like a sick man whose only hope is a difficult operation. We cannot be sure it will save us, but since it will lead to health if anything will, it is rational to submit to the operation.

But is it our only hope?

I have used the example of the operation precisely because many people *reject* its logic. 'What about miracles?' they say. Or faith-healing? Or spontaneous remission? What if your *diagnosis* is in error? In the extreme case of the Christian Scientists, it will be claimed that not only is the method of faith a practical alternative to the operation, it is also morally obligatory. These considerations force us to consider the analogous possibility in epistemology: Might there be alternatives to the Rule of Induction?

Reichenbach considers the possibility that an alternative might exist. As the simplest and most forceful example, he talks about a Seer who foretells the future with great accuracy. This certainly seems to be a better way than induction. But, says Reichenbach, if the Seer continues to outdo us with great consistency, induction will take note of that fact and will tell us to listen to the Seer. It will, in effect, discover a new regularity in the world that will serve as a valuable guide to the future. Thus, again, induction will have led to success where success is possible.

Now I think Reichenbach's argument is powerful and very suggestive, but I do not think, as he did, that it is a decisive demonstration that induction *will* lead to success wherever success is possible.

To expand a bit on the above possibility, let us assume that the human brain is inherently capable of understanding and interacting with the world in either of two mutually exclusive ways. The first can be our Western, scientific, approach of understanding the world by induction and directly influencing only our bodies. The second is suggested by the type of thing we categorize as mystical, magical, or intuitive, but we will assume it possesses a far greater efficacy than generally imagined, so that it is possible for an ordinary person

196

who follows this course of development to arrive at a direct, non-inductive, understanding of the universe and directly to affect any physical thing (telekinesis, miracles) or mental entity (telepathy, 'voices') in the universe.

We now assume that the 'branching' of the alternative courses of development occurs sometime in late childhood, when a person has a fair acquaintance with the concepts of 'evidence' and 'scientific procedure' and also some familiarity with ghost stories, ESP, miracles, and 'acting on faith and intuition.' Those few who choose the non-inductive route (or just drift into it) we can imagine as constituting a marvelously happy and powerful but *invisible* sub-culture. They have chosen not only not to inform us of their methods, but actually to *prevent* our acquiring any significant evidence of their blissful existence (a few ghost stories, flying saucers, and religious fanatics are harmless, and may furnish an acceptable level of 'recruitment,' but no 'scientific' evidence is ever allowed).

Thus we have a situation where induction is not the optimal procedure for dealing with the world. It can even be the *worst* procedure if The Others intervene to ward off the really calamitous results of irrational activity during the time when fools and children are 'setting' in the non-inductive mode of development; for a time they *will* be less successful than we (so that our evidence is always that they are wrong) but after they survive that period of unhappiness, they so far outstrip us that we can say that the worst thing that can happen to a kid is for him to get hooked on the mild pleasures of induction at the expense of his eternal happiness.

Finally, let us suppose that the world is such that it is *not* characterized by the widespread existence of limits to empirical frequencies. Then most of our inductive knowledge will gradually break down, the hopes of scientific optimism will never be realized, and most of mankind will forever suffer poverty, hunger, and ignorance. The Others continue to succeed with their non-linear projections of the future; inductivists continue to fail with the Straight Rule.

In this fictitious world of ours there are two key frustrations to ordinary induction. The first challenges Reichenbach's assumption that prediction of an empirical series is only possible if that series has a limit. Interestingly enough, Reichenbach immediately gives up this presumption and withdraws to his second line of defense. If there were a non-inductive method that could lead to success in such cases, he argues, the use of such a method would itself generate a success-frequency which could be detected by induction and subsequently adopted by the inductivist. Hence his famous claim that induction will lead to success in prediction if success is possible.

But all that needs to be done to refute this claim is to build into our worst-case (diagonal) argument a provision which frustrates the one induction which could tell the inductivist to abandon induction. In our argument, this second frustration is the fact that the Seers are *clandestine*, so that success is available to each individual but that fact is never revealed by induction. A more traditional situation might have Descartes's demon gleefully falsifying the evidence every time an inductivist gets close to understanding. Or a fundamentalist Christian's God might build into his universe false evidence of the value of science, in order to test the faith of believers by tempting them with never-quite-fulfilled promises of scientific success.

Any number of worlds, in fact, can be imagined where it is not true that the Straight Rule of Induction will lead to success in prediction if success is possible. The only essential elements are (1) limits don't generally exist, so the Straight Rule is not generally successful, (2) an alternative method exists which is successful, and (3) some feature of the world prevents induction from discovering the alternative method.

This argument might be viewed as an extension of Putnam's diagonal argument against best-possible inductive mechanisms (although this is not in fact how the argument developed).

Putnam, you will recall, argued that given any c-function one could always design a world and a second c-function which does a better job in that world than did the original. I am arguing that given this particular inductive method (the Straight Rule), one can always design a world in which some alternative (perhaps non-inductive) method works while this one does not and also make that fact undetectable by the Straight Rule. If even one such world can be constructed – no matter how malicious and *ad hoc* our procedure – then we have refuted Reichenbach's claim that the Straight Rule will always lead to success if success is possible.

Reichenbach is correct that we do not *appear* to live in such a world, but, in the first place, the evidence may be *rigged*, as above, and, in the second place, it is only a contingent truth if we do not. It is certainly not logically true that induction will lead us to a successful means of dealing with the world if such a means exists.

It may not even lead to *any* success in predicting the future, for if They allow only random success for induction and accept for Initiation only those who, for three years or so, employ some unproductive, non-inductive method which *seems* to yield them no success at any time, induction will lead to no general success and one major error in predicting the future, and will continue to support its own employment, or at least not reject it (because the 'evidence' is that it is at least as good as the 'unproductive' method). It is therefore logically possible that man can have free will and the ability to affect his future and a way to achieve success in predicting that future, while induction is not that way.

We must conclude that the pragmatic justification of induction fails in its attempt to *demonstrate* that the inductive method must lead to success, if success is attainable.

What we can say of induction is that it *seems* to work well, and our understanding of it shows that it is so constructed as to be successful if the nature of the world is not radically different, in peculiar ways, from what we take it to be. While this is not enough to insure that induction is *necessarily effective*, it may well be enough to convince us of the rationality of acting on the Rule of Induction. As Reichenbach says:[90]

> a man's actions can be guided only by his knowledge of the world, and not by unknown features of the world; if we have to decide whether his actions are reasonable we have to ask whether they are reasonable in relation to what he knows.

Surely we can say that, on the basis of what we know, the Rule of Induction has a strong advantage over any other practical methodology we might employ. This plea for the reasonableness of induction will not meet the standards most impose on any 'justification' of induction, but it will remain true that most reasonable people act inductively.

10   CHIEF CRITICISMS OF RF THEORIES

We will begin by discussing that cluster of related criticisms concerning what might be called the *epistemological difficulties* of RF theories. Three major ideas seem to be involved in these criticisms:

1 RF probabilities can never be known.
2 RF probabilities can never be known to exist.
3 A putative RF probability value is neither confirmable nor disconfirmable.

These ideas are interrelated and not really distinct, but it will suit our analytic purpose to attempt to discuss each in turn.

To begin with the first one, there is an obvious sense in which an RF probability statement would be just as unknowable as any other (ordinary) empirical description of the world. RF theorists accept this lack of certainty and even delight in it. It shows, they say, that RF probability statements are just like statements about an object's mass (a favorite comparison). Such statements are inherently uncertain *because* they describe the real world, but this lack of certainty is no bar to the value of scientific statements nor even to their accuracy. We admit that our value for the mass of the moon *may* be wrong, but we think it is *better* than earlier values, that it is very likely correct to within a tolerable limit of accuracy, and that it is pragmatically sound and eminently reasonable to act on such values even if there is a non-vanishing chance that they may be wrong.[91]

We may take it that the fact that RF statements cannot be known with certainty *merely because they are empirical* is recognized and accepted by most RF theorists but is not considered an *objection* any more than it is an objection to any empirical science. If this were the extent of the criticism, RF theorists would seem justified in shrugging it off as the whining of metaphysicians in search of apodictic certainty. But this is *not* the extent of the criticism. While the critics do, of course, believe that RF statements are subject to the generalized uncertainty of all empirical statements, they further contend that RF statements are subject to a qualitatively different and far more debilitating uncertainty which derives from their peculiar nature as limit statements.

Consider the difference between the statement that the mass of the Moon is X and the statement that the probability that a star will be a red giant is P. The first statement is justified by an accumulation of measurements using the best instruments we can construct within the context of the best theories of reality we have at hand. We are prepared to admit, on the one hand, that there might have been measurement error, either in our instruments or in our use of them, or, on the other hand, that a theoretical advance such as Einstein's might lead us to revise our notions about *what mass is* and therefore about what the moon's mass is. In these two principal ways, I think, we are prepared to admit that our value, X, might be mistaken, even beyond the prescribed limits of accuracy.[92] What we are *not* prepared to admit is that our measurements might be exactly correct and our theory impeccably reliable and we might *still* have a significantly wrong value.

But now consider the statement that the probability that a star will be a red giant is P. This statement is likewise based on our measurements of the world – it is therefore subject to the same doubts concerning measurement error and theoretical adequacy as is the first. Let us assume that no difficulty exists in either category, that we have correctly and precisely examined 100,000 stars of which precisely 10,000 have been correctly identified as red giants. We therefore assert (or, as Reichenbach prefers, 'posit') that $P = 0.10$.

Now suppose we continue to observe new stars, either those we 'haven't got around to' before or more distant ones newly apparent to improved telescopes. We *might* find that the relative frequency of red giants in the newer observations is greater than that in the old – say 0.20. If the trend continues in this direction, after an additional 100,000 observations we will have discovered 20,000 new red giants. The *overall* relative frequency of observed red giants has thus become 0.15. If we now agree that *this* value is experimentally and theoretically correct, we are forced to admit that our earlier value for P contained an error of 50 per cent *even though no experimental or theoretical error was involved*. It is inconceivable (save in the sense that Descartes' demon is conceivable) that we would ever come to admit that our present belief about the mass of the moon is in error by a factor of 0.5 even though no theoretical or observational error had been committed.[93]

The difference here is that properties such as the mass of a body or its color or velocity are contingent properties of *individuals*, while RF probability values are contingent properties of infinite *sequences* of individuals (or events). By 'contingent' I mean that we have no way of knowing in advance (a priori) just what value the property has (either qualitatively or quantitatively). Leaving aside predictions based on cosmological theories, let us agree that the mass of the moon and the frequency of red giant stars are alike contingent in that we can have no good idea of their values until we *look and see*.

Now to 'look and see' what the mass of the moon is, we do such things as measure its orbit, its influence on space probes, its perturbations by other heavenly bodies, the readings of spring-scale weights of standard masses on its surface, etc. In a sense it is possible to continue this sequence of measurements indefinitely, averaging the results anew each time, and therefore 'refining' the value for ever and ever.[94] But if such future observations give significantly different values for X they are held to be *inconsistent* with the earlier measurements, and one or the other (or both) must be explained away as resulting from measurement or theoretical error. In short, a reasonable and finite number of measurements can *establish* the value of ordinary physical properties so convincingly that any significant revision must include an acknowledgment of previous error. (How many times must we look at *one* star, for example, before we take as established that *it* is a red giant?)

But RF probabilities are not properties of individuals. They purport to describe infinite sequences of individuals. How can we hope to justify or establish a statement like this?

Well, first we 'look and see' what property individual A has. Then we look and see what property individual B has, etc., etc. Now each examination of an *individual* includes all and only the epistemological difficulties of establishing what property an individual possesses (e.g., the mass of the moon, or whether *this* star is a red giant). But we can never examine *every* member of an infinite collection. Therefore, at any given time our value for P will be based on a sub-sequence of observations which may contain a great number of individuals but which is always finite. *But the value of a relative frequency established by any finite sub-sequence is compatible with any value whatsoever of the limit of the relative frequency in the entire infinite sequence.*

This is a mathematical fact which is apparent to anyone who reflects on the nature of infinite sequences.[95] In our example about red giants we saw how the value of P could be increased 50 per cent by doubling the number of observations. Obviously *that* could likewise be increased (or decreased) significantly by an additional 500,000 observations. And if we consider what is *ex hypothesi* a contingent property, it is clear that the 'final' value of the relative frequency *could* be anything at all, no matter what we have observed in a relatively insignificant sample – and any finite sample of an infinite collection is always relatively insignificant!

202

We see, then, wherein consists the difference between RF probabilities and 'ordinary' physical properties. In the latter case we admit that we might revise our value – but only if we are convinced that there is some *error* in present measurements. In the case of an infinite sequence, however, we might have done everything exactly right up to this point and *still* be way off the mark for the 'true' value. An RF statement therefore contains an ineliminable uncertainty which is *not* epistemologically similar to that of ordinary statements of the physical sciences and should not be dismissed as trivial.

Now we pass on to the second claim of the critics: that RF probability values cannot even be known to exist. We will discuss this point only briefly because it is unintelligible to those who don't understand limits and infinite series and obvious to those who do.

Briefly, a series is said to have a *limit*, L, if for any small value, d, there exists a point in the series identified by the integer N after which the value of the series never deviates from L by more than d, no matter how much further in the series one goes.

Take for example the mathematical series specified by

$$S_N = 1 + \frac{1}{N+1}$$

the first members of which are

$$2, 1\,1/2, 1\,1/3, 1\,1/4, 1\,1/5, 1\,1/6, \ldots$$

and the graph of which is

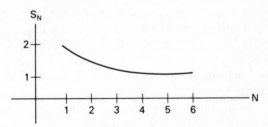

This series is said to approach 1 as a limit because if you specify any small but finite value d, I can identify a point in the series after which $|1 - S_N|$ is always less than d. If you say $d = 0.55$, then $N = 1$. For $d = 0.23$, $N = 4$, and so on. In this case, we can know that the series approaches a limit because the series is *intensionally given* (or defined mathematically).[96] The rules of mathematics suffice to guarantee the properties of the series, including the existence of a limit. Therefore we can be sure of the existence of the limit and even of its value.

But RF probability statements deal with *empirical* sequences which can only be given extensionally (that is, by specifying which individuals are members but not what their properties are), not intensionally. For example, the series of stars (or observations of stars) is specified in a way that doesn't tell us whether a given star will be a red giant, or even how many will be. If the relative frequency of red giants is considered as a sequence, therefore, there is no a priori method of determining if the sequence even *has* a limit, much less what it will be.

In fact, the only sequences which we know *do* have limits are those which are defined in a priori mathematics. But, as C. I. Lewis says, the frequentists can take no comfort from the existence of these limits to infinite mathematical series. For even if an empirical series has closely approximated such a mathematical series over an initial stretch, it is under no compulsion to do so in the future, and certainly not through the entirety of an infinite future extension. The unwarranted use of mathematical series in the place of empirical ones is a 'besetting fallacy' of the frequentists.[97]

Furthermore, there exist mathematical series of which we know that they do *not* have limits (all divergent series, for example). What if empirical series really imitate these series? In our astronomical example, for instance, it may be that the proportion of red giant stars in the universe is subject to temporal variations of such a nature that their relative frequency of appearance does not approach a limit. If this is so, then we are not just subject to a quantitative error when we say 'The probability that a star is a red giant is 0.09' – we are literally talking nonsense. For if the series does not approach a limit in the long run, there is never a probability of the RF type *at any point* of the series. The statement is very likely inaccurate (since I just made it up); but could anyone convince a working astronomer that it might even be meaningless and there might be *no* probability that a star is a red giant? What would it mean to say that there is *no* probability that a star is a red giant when we know that many *are* red giants? This would clearly be disanalogous to the question of what is the mass of the moon, where we are prepared to believe that we may have the wrong value, but deny that there might be *no* value.

Reichenbach responds to this criticism by arguing that at least some series must have limits if the world is to be predictable in the long run, and our only *hope* of success lies in acting as if the world is predictable. But of course short-range predictability could give us all the success we finite humans could hope for without the existence of limits, and merely 'next-case' predictability – making possible eduction to particulars, but not induction to laws – would serve well enough to keep the community going.

Returning now to the question of what the value of a limit is – assuming it does exist – our third version of the epistemological difficulty holds that if *P* is a statement asserting that a certain RF probability has a certain value, then *P* is both unverifiable and unfalsifiable. This is very nearly a restatement of the first formulation – that the limit cannot be known – but our emphasis here will be somewhat different.

First, *P* is unverifiable. This point is made by many critics of RF theory,[98] and is, indeed, admitted by Reichenbach:[99]

> With respect to intensionally given probability sequences of an infinite length, statements expressing a frequency interpretation have the usual meaning of mathematical all-statements and existential statements; like the latter, they are strictly verifiable. With respect to extensionally given probability sequences of an infinite length, however, statements expressing a frequency interpretation are not verifiable.

These two statement-types parallel the distinction we made earlier[100] between PT (theoretical) and PA (applied) statements. The former class is verifiable a priori, by the laws of logic and mathematics, but they say nothing about the world. All statements about probability values in the world are PA statements (according to me) and given extensionally only (according to Reichenbach) and therefore never completely verifiable.

If an empiricist adhered to the strict interpretation of the Logical Positivists' Verification Theory of Meaning, he would be forced to abandon RF theory as being meaningless because not completely verifiable in experience. But most empiricists have relinquished the strict verification theory in favor of some more tolerant method, such as Carnap's Confirmability Principle. Perhaps we can do the same with RF statements. That is to say, even though we admit it is impossible completely to verify *P* (since it refers to a class stretching infinitely far into the future), isn't it possible to collect enough evidence to make *P* probable? No, says Lewis,[101] because

> If we apply this same [RF] interpretation to the *probability* that '*P*' has probability *a/b*, then it must appear that there is no certainty but only a probability that "'*P*' has probability *a/b*" is probable. And so on. Thus when confronted with the general problem how we are to elicit or express the cognitive status of beliefs which have some justification but are less than completely certain, we find that the empirical interpretation of their probability would not provide a solution but only the beginning of a perpetual stutter.

If we had an independent, a priori, inductive logic such as Carnap's, we might be able to say 'It is (AP) probable that *P* (RF statement) is true.' But if we restrict ourselves, as Reichenbach does, to the RF interpretation of probability in all circumstances, it is not clear that there is any way in which we can justifiably argue that *P* itself is probable.[102]

Let me re-emphasize this point, because some proponents of the RF theory have tended to shrug off this criticism as being excessive attention to the relatively cheap and easy verificationist argument and as being solved or eliminated by turning to confirmation theory or rules of acceptability. But such epistemological maneuvers do not solve the theoretical problem at all (although they may be useful to the social scientist or applied mathematician who is seeking practical guidance about how to judge probabilities.) The reason is that confirmation theories or acceptance rules are themselves probabilistic in nature; they don't guarantee 'success' each time but only rationality over the long haul. Now if these rules are to be justified theoretically (as opposed to remaining practical rules of thumb), they must be treated in the context of some theory of probability or induction. If that meta-theory is a priori in nature, then Carnap is right in claiming that AP theory is conceptually more basic than RF theory (*contra* Reichenbach), but if the meta-theory is itself an RF theory, then we raise at the second level the very same epistemological difficulties we encountered at the first level and we have Lewis's 'perpetual stutter.' So invoking confirmability rather than verifiability, while it improves the practical situation (by allowing action on less than conclusive evidence, e.g.), does not alter the theoretical situation one bit.

Finally, then, suppose that we abandon confirmability as well and try the falsifiability criterion of meaning.[103] Perhaps P shares the property of most universal generalizations of being never verifiable but at least conclusively falsifiable through modus tollens.[104]

Unfortunately, this is not the case. The attractiveness of falsification consists in the fact that a single example suffices to falsify a generalization like 'All stars are red giants.' We could never in principle find enough red giants to guarantee the truth of this statement, even if every star we examined for a thousand years turned out to be a red giant. But if we found *just one* star that was *not* a red giant, we could know conclusively that the general law was false.

But now suppose we consider P – 'The probability that a star is a red giant is 0.09.' Our earlier discussion purported to show that nothing could ever conclusively verify this statement (because a stretch of ever so many observations exhibiting a relative frequency of exactly 0.09 is logically compatible with any value whatsoever for the actual limit, if there even *is* a limit). A strictly parallel argument will show that it is also the case that no finite sample can ever conclusively falsify an RF probability statement (since it is compatible with any value whatsoever...etc.).

Consider an analogy. Suppose the president of our university claims that we are in compliance with federal anti-discrimination regulations because 0.09 of our students are black. 'But,' says a visiting Congressman, 'I just saw two of your students in the hall and they were both white.' What can we make of this besides gross stupidity? Surely such a small sample can't be seriously intended as a falsification!

Now suppose I assert our astronomical statement P and the chairman of the Astronomy Department gently but firmly disagrees. 'After all,' he says, 'we have examined over 3 million stars and have consistently found 0.18 of them to be red giants.'

As it happens, $3,000,000/\infty$ is a smaller fraction than $2/18,000$. Therefore the astronomer's sample is proportionately smaller than the Congressman's! If it would be absurd to consider the latter to be a falsification, on what grounds could we claim that the former is?

I think there are two chief reasons why we intuitively reject one 'falsification' and tend to accept the other. First, we think that the Congressman has failed to make any effort to examine readily available data that would greatly increase his sample size, and presumably I, being no astronomer, have done the same. Second, we reasonably assume that both the astronomer and the university president have much larger sample sizes than their antagonists – thus falsification attempts fail against larger samples and succeed against smaller ones.

To vitiate these intuitive epistemological responses, suppose my rejoinder to the astronomer is 'Oh yes, I know that, but my brother is in charge of processing data from the new orbiting astronomical observatory on his company's giant scientific computer and he just told me that the newest and most remarkable result is that by doubling our sphere of observation we have found that more distant stars exhibit a considerably different distribution along the stellar sequence. In other words, I have examined *six* million stars (by proxy) and found only 9 per cent to be red giants.'

Now do you think that the astronomer's true statement about his research constitutes a *falsification* of P? Or even that *he* will continue to think that it is? The reason we were at first inclined to accept it as a refutation of P is that it seemed to embody the best scientific opinion based on the greatest amount of data. The fact that we are willing to change our opinion under the new circumstances clearly shows that the astronomer's *evidence* does not *contradict* P at all – the two are perfectly consistent. Therefore 'the best available astronomical evidence' can never *logically falsify* P.

But *every* RF statement concerns an infinite population of which only a small portion has been examined. Therefore no RF statement can ever be conclusively falsified by *any* (finite) amount of evidence.

We have reached a curious epistemological position. Scientists and empiricists embrace RF probability theory because they want to tie probability to the real world and make probability statements just like other scientific statements. But if these people also embrace the traditional empiricist position that all meaningful statements must be either (a) a priori truths of language, logic, and mathematics, or (b) empirically verifiable/confirmable/falsifiable, then it seems to follow that RF statements are *meaningless*, while AP probability statements may be meaningful, if it can be shown that they properly fit into the first category. This is the heart of the epistemological difficulty of RF probability theory.

After the epistemological argument, perhaps the most common objection to RF theory is that it changes, ignores, distorts or limits the meaning of the word 'probability.' Keynes says:[105]

208

It is the obvious, as well as the correct, criticism of such a theory, that the identification of probability with statistical frequency is a very grave departure from the established use of words; for it clearly excludes a great number of judgements which are generally believed to deal with probability.

This objection is barely open to Classical theorists, since RF theory can deal with all the same types of events that they can, plus others besides (biased dice, death probabilities, etc.). The only real possibility for Classical criticism is that RF theorists have eliminated probabilities for single events – von Mises explicitly did so, while Reichenbach attempts to preserve a 'fictitious' meaning in such cases. But the most important uses of Classical probability (speaking as a poker player) are just such statements of single-case probability ('The probability of filling this inside straight is 4/47 while the probability of improving my pair is 18/47 so I'll draw to the pair'). After all, a gambler doesn't really care about general statistical relations. He wants to know what his odds are *this time*, and (some) RF theories refuse to tell him that. They therefore fail to capture one of the most important uses of 'probability' by Classical standards.

A priori (AP) theorists can press this claim with much greater force, since their theories account for the same uses as the Classical theory plus the uses of the RF theory plus a great deal more. In Keynes's view, every time we use a word like 'probability' or 'probably' or 'likely', we are referring to a simple property open to our intuition and dealt with by the theory of probability. According to Carnap, every such statement is related to the evidence for it by logico-mathematico-linguistic relations which (usually) allow us to compute its confirmation by that evidence. Thus both AP theories deal with *all* uses of the word 'probability.' RF theory, however, ignores not just single-event probabilities but also all evidentiary statements of less than certainty ('On the evidence, he is probably guilty'), all speculation about the future ('The Cardinals will probably win the pennant'), or about unknown facts ('Probably 75 per cent of all Russians are religious'), and, in general, any case in which it is not possible to speak directly about one infinite class as a subset of another.

Many people who raise this objection do not intend to attack the value of appealing to frequencies in some cases. They wish only to point out that success in the limited area of scientific and actuarial cases is no good reason for RF theory to claim to be the whole of probability theory.

Another 'criticism' that is often raised against RF theory is more in the nature of a *tu quoque* response to RF charges of relativity in AP theory: the criticism, namely, that RF theory makes all probabilities relative to a reference class. When RF theorists claim that they are not interested in what the probability of X is relative to some evidence set or other but only the probability of X, it is natural and appropriate for the apriorist to reply ('So's your old lady') 'But your theory likewise does not specify absolute, independent probabilities, only probabilities relative to some reference class or other. So get off our backs!'

Now I think this 'criticism' both does and does not succeed, and I think that only a little careful analysis will be required to distinguish its successes from its failures.

First, it succeeds in knocking down any RF theorist who thinks that his theory does, while the AP theory does not, establish real, absolute values for 'the probability of X independent of other considerations.' It is just not possible for an RF theorist to give the probability of being an A as such, he can only give the probability that a B is an A, or the probability of A in a specified reference class (collective). The very definition of RF probability requires that there be a denominator (reference) class as well as a numerator (quaesitum) class in order to specify a frequency and, thereby, a probability. Any RF theorist who fails to realize this is subject to a grievous error. (It is not, incidentally, an error committed by either von Mises or Reichenbach, both of whom recognized this dependence.)

Second, this 'criticism' does not just point out a *feature* of RF probability, it recognizes a genuine *difficulty*. In our discussion of Reichenbach's theory we have already dealt with the problems of selecting reference classes and sample sizes.[106] (Briefly, if 50 per cent of all college graduates but 20 per cent of all churchgoers drink alcohol, what is the probability that Sam Jones, a churchgoing college graduate, drinks?) This selection is *important*, because the probability value changes with the nature and size and even the order of the reference class.[107] And it is *difficult* because there is no theoretical rule which determines the selection in an unambiguous manner. (I judge Reichenbach's admonition always to select the smallest class for which reliable statistics exist to be a helpful rule of thumb which cannot cover all situations; see above, pp. 165–6.) To this extent, the criticism does indeed identify a *problem* in RF theory.

But my third point is that this problem is not peculiar to RF probability theory, since it occurs in some guise in each of the other two theories which we have discussed, as well. In the Classical theory, it amounts to the problem of specifying equiprobable alternatives. In AP theories, it is the problem of specifying the hypothesis and the evidence and selecting the correct language for doing so.

In the example of the churchgoing college graduate, Classical theory is almost useless. If we assumed that each predicate was an independent Bernoullian property with Yes or No equally likely, then the chance of being a college graduate is 1/2, the chance of being a churchgoer is 1/2, and the chance of being a drinker is 1/2. Thus, by multiplication of probabilities, the probability that anyone is a religious drinker with a degree is 1/8. But if we know or discover that Sam Jones is a churchgoing college graduate, the probability that *he* is a drinker reverts to 1/2.

A better alternative is to assume that churchgoing and being a college graduate are independent properties with the stated probabilities and then to use the method of the composition of chances to 'average' the two probabilities so that $\dfrac{5/10 + 2/10}{2} = 7/20$ is the probability that Sam Jones is a drinker.

The pure or 'null' AP probability (Carnap) that Jones is a drinker is constructed exactly like the first Classical method and likewise leads to the probability of 1/2. If the evidence is expanded to include the known facts and the independence assumption is otherwise retained, the AP method again agrees with the Classical that 7/20 of all churchgoing graduates are drinkers.

Now these methods have proceeded by simplifying assumptions which allow them to attain their definite numerical solutions. But the RF theorist could also solve the problem *if* he made these assumptions. That is, if one series of churchgoers containing 20 per cent drinkers is merged equally with another series of college graduates containing 50 per cent drinkers to give a third series which is 50 per cent college graduates and 50 per cent churchgoers, the final series will contain 35 per cent (7/20) drinkers. Thus the same assumptions lead to the same solution. The difference is that the RF theorist hopes to avoid such a priori assumptions while the AP and Classical theorists are more likely to accept them. The important point is that the answer cannot be obtained in either case without these assumptions and it *can* be attained in either case if they are accepted.

It might be argued that the other theories are superior to the RF theory because they build the assumptions into the theory and thereby *directly* sanction the answer while the RF theorist can only obtain it by making *ad hoc* additions to *his* theory. This is certainly a *difference* between the theories, but it is clearly debatable whether it is an *advantage* for one or the other. It is perfectly possible for an RF theorist to respond that in this instance – as in most – the independence assumption is more likely to be false and misleading than advantageous.

It is not, of course, obvious just *how* the assumption might go wrong. A pro-religion, pro-education observer might argue that educated believers more clearly understand and more reliably obey the injunctions of their religion so that their lapses will be less frequent than those of their ignorant brothers, amounting to a total of perhaps 10 per cent drinkers. Anti-religious/pro-education observers might argue that education leads one away from religion (thereby 'liberating' him) so that those educated church members are there for social rather than religious reasons and will reflect the drinking probability of the unchurched, 50 per cent. A real cynic might hold that any educated person hypocritical enough to go to church probably possesses other vices as well and might therefore be *more* likely to drink than his educated unchurched peers. The possible arguments and combinations are numberless. There is no obvious way of deciding *which* of them is correct, but they certainly conspire to make it seem unlikely that there is *no* interrelation between the properties. Therefore, the RF theorist might argue, we are much better advised to withhold judgment until we can examine the facts than to acquiesce in an automatic and simple-minded assumption of independence.

I do not claim to have settled this dispute one way or the other. I do hope I have made clear why the RF theorist's problem of identifying a reference class is not necessarily either unique or debilitating to his theory.

My third point is that it is not the case that RF and AP theories are 'relative' in the same way. The usual complaint about AP theories is that they make probabilities relative *to human knowledge*. Now this may or may not be a sound objection (the rejoinder is that probability, like deductive logic, is an objective relation between propositions which obtains completely independently of psychological vagaries). But it is certainly different from saying that RF probabilities are always relative to empirical and objective reference classes. *If* the RF criticism of the apriorist is correct, then its teeth are not drawn by showing that the RF theory has a *different kind* of relativity. We must then take further thought to determine which – if either – form of relativity is discrediting.

To sum up, RF theories do indeed make probabilities relative to reference classes, but it is not at all obvious that this is a peculiar or incapacitating property of those theories; it might very well be that such 'relativity' is an essential part of our concept of probability and not be eliminated by any good theory.

Our next criticism of relative frequency (RF) theories is that they are irrelevant to any actual events. This contention is somewhat ironic, since one of the arguments for moving from a priori (AP) to RF theories has been that the latter are more 'scientific', realistic, empirical, and tied to the real world than the idealistic, formal systems of the AP theorists. Now the argument seems to be reversed.

The major reason for holding that RF theories are irrelevant to real world events is that they define probability in terms of infinite series and *no such infinite series exist* (at least not within human experience). All real events are finite and limited. We are never concerned with questions like 'What proportion of Sevens will occur in an infinite number of rolls of these dice?' but rather with 'What are the chances of throwing a Seven this time (or before I make my point)?' or, at most, 'If I always bet on Seven (throughout my life) what are my odds of winning?'

In my discussion of single events I argued that von Mises' theory is clearly not applicable to such events and that Reichenbach can only apply his theory by an extension which clearly violates his original definition of 'probability'. I will not repeat myself in detail at this point, but the reader should recall the gist: if probability *is* the limit of a frequency in an infinite population then probability is *not* the chance or likelihood that *this* event (or finite series) will have a certain property. The definitions simply do not mean the same. Since we never experience infinite series, it follows that RF probabilities are never directly relevant to human experience.[108]

Why have the 'empirical' RF theorists settled on a concept which by definition is never apprehensible in human experience? Carnap and Kneale, in slightly varying ways, have argued that RF theory involves a confusion. Kneale has argued that, like the 'constancy' theory of natural law, RF theory confuses the evidence with what it evidences.[109] Carnap suggests a confusion between a frequency and the estimate of a frequency.[110]

Kneale's point is this. Suppose there is a 'probability' (in the ordinary, intuitive sense) of 1/6 that a die will show a Five. One of the reasonably expected consequences of this is that is that a fairly long series of rolls with that die will consist of approximately one-sixth Fives (Bernoulli's Theorem and the various Laws of Large Numbers formalize this expectation).

Now suppose we observe a long series of rolls and Fives instead appear two-thirds of the time. Surprised, we try again, and again, and each time the frequency of Fives centers around 2/3. This seems to give us evidence that the probability of a Five is not 1/6 but 2/3. (Bayes's Theorem is approximately a measure of how good this evidence is. That it cannot be conclusive should be apparent from the earlier discussion of the non-confirmability and non-disconfirm-ability of RF probabilities.) After *enough* Fives have shown in this manner, nearly everyone would agree that the die is biased (especially if the discussion came not in the context of competing theories of probability but during a high-rolling crap game where they were expected to 'put their money where their mouth is').

Thus a well-established frequency is good evidence that a probability does not have a certain (originally expected) value and good evidence that it does have a certain (different) value. It is *such* good evidence that in practical matters it quickly outweighs and overrides 'merely theoretical' or 'a priori' considerations. Hence its popularity with scientists and businessmen. (It seldom interests gamblers since their devices are intentionally constructed *to conform* to the much simpler Laplacean system of equiprobable alternatives.)

Indeed, in cases where new properties or combinations of properties are being investigated we may have no good way of calculating probabilities in advance. All we can do is 'look and see' what the frequency is. Repeated experiences like these have shown that many probabilities can be established by investigating frequencies which cannot be established by counting equiprobable cases a priori and cannot even be incorporated in such a schema after the facts are known. Thus the classical definition was found to be inadequate in many situations and was abandoned. All that was left was the frequency which had demonstrated its importance as evidence. In many cases, frequency was the only criterion which worked, and it seemed to work in most cases. What could be simpler than to identify the probability and the frequency?

This identification may very well be an instance of the Wittgensteinian mistake of confusing the criterion with the concept,[111] but it will serve just as well to treat it in Kneale's terms as confusing the evidence with what it evidences.

Remember, we started with the assumption that the die had a certain probability of showing a Five. The frequency was important as evidence that it was really a different probability. This was its only importance! Suppose we discovered that the evidence had been faked, or mistakenly reported, or caused by some eliminable outside force. Would we not say that the frequency was 'caused by' something else and not good evidence of the probability at all? And if a 'good die' were rolled in such peculiar circumstances throughout its entire (fairly short) life, and then destroyed, would we not say that the (simple, natural, initial) probability of throwing a Five with that die was 1/6 even though the relative frequency of Fives in fact had turned out to be 2/3? This is because the frequency is *only* evidence for the probability. Normally it is very good evidence, but sometimes it can be 'explained away.' But if it can *ever* be 'explained away' it cannot *be* the probability. Hence Kneale's argument that RF theories confuse the evidence (the frequencey) with what it evidences (the probability).

Carnap's complaint is somewhat different. He accepts the identification of the probability with the relative frequency in certain cases (he calls this 'probability$_2$'). But he protests when Reichenbach goes on to apply this concept to judging the 'weight' (probability$_1$, confirmation) of an unknown proposition. Here Carnap agrees that it can be useful to speak of the 'truth-frequency' of the proposition (i.e., the percent of its normal applications when it will be true). But it is a mistake to identify the probability with the *actual* truth-frequency, since this is *ex hypothesi* unknown to us and can be of no help in our present efforts. What we actually want is an *estimate* of that frequency, based on our best available evidence. That is, we are seeking guidance as to how we should act with respect to a certain proposition. If we *could know* how often it will be true throughout the infinite future, that would surely be a great help. But for contingent propositions such knowledge is clearly beyond us – the best we can do is to appraise the evidence and conscientiously *guess* at the truth-frequency.

This leads to two related criticisms: that RF probability does not change with experience and that it is unsuited as a 'guide of life.'

The reader may be surprised that I cite 'learning from experience' as a difficulty with RF theory, since its advocates have long claimed that one of its chief virtues is the way it is 'tied to the real world' rather than based on 'a priori rationalism.' The difficulty is that RF probabilities, while tied to the real world, are not tied to *human experience* of the real world.

For example, the probability that a star is a red giant is equal by definition to the relative frequency with which such giants appear among stars in the entire history of the universe. This number, if it exists, has one and only one value which is eternally constant. It is an objective fact which is unaffected by human knowledge, beliefs, hopes, or experience.

The peculiar consequence is that it is always false to say such things as, 'The probability that the Red Sox will win the pennant is greatly diminished by their loss of Fred Lynn,' since it is either false that there *is* such a probability or, if there is, it can neither be increased nor diminished by anything at all.

Such eternal changelessness may not confuse scientists, since they are generally willing to admit that when they adjust probabilities they are changing only an estimate or accepted value rather than asserting a change in the objective value. (This is likewise true of most other scientific changes, like changes in the value of 'the mass of the moon.' A disanalogy: There might be a *real* change in the mass of the moon, if it were hit by a sizeable planetoid, but it is impossible for there to be a real change in the probability of such a collision.) But in ordinary life we often talk of increased probabilities, better odds, 'more likely than before,' etc. If the RF theory is taken literally, all such talk must be abandoned as false, since probabilities never change.

Eternal unchangeability and unaccessibility are also part of the reason that Carnap claims RF theory is unsuitable as a 'guide of life.' This point was discussed in some detail in the section of the AP chapter which dealt with the rationality of probability behavior. The fundamental argument is that one cannot use an unknown quantity as a guide, and RF probabilities are either completely unknown or at least 'less knowable' than the computed values of AP probabilities.

A similar argument can of course be made to show that RF probability cannot serve as an inductive logic.

A reverse twist of this criticism is that to the extent that the RF theory embodies truth, it *is* logical, analytic, a priori. Specifically, the axioms and theorems of the calculus, which alone are free from dispute as part of the theory, are developed by von Mises unwittingly and by Reichenbach consciously as a formal axiomatic system – a part of pure mathematics, not of science.[112]

Finally, I would like to remind the reader that the strongest objections to RF theory are its epistemological difficulties and its inapplicability to many uses of 'Probability'. But not to be overlooked is Keynes's 'reputed rejoinder concerning the practical value of the long run justification,' that is, 'in the long run we shall all be dead.'[113]

11 CHIEF VIRTUES OF RELATIVE FREQUENCY THEORIES

I believe that there were two main reasons for the considerable success the RF theory had in gaining adherents – one was psychological and the other practical.

The psychological reason was the RF theory's claim to be 'scientific'. Clearly it was better suited to the tenor of modern times to appeal to 'empirical evidence' and 'scientific observation' in establishing initial probabilities than to use the old AP method of laws of rationality and intuitive equiprobability. Indeed, for empirically oriented people, it was difficult to resist the appeal of the theory which located probabilities in the objects themselves and claimed they were discoverable by the normal methods of science – especially when the only clear alternative theories made probability a feature of our attitude toward things, or, at best, a logical feature of our descriptions of things. In the context of the 'Rise of Scientific Philosophy'[114] it seemed natural and appropriate to adopt a 'scientific' theory of probability.

But of course this psychological appeal was not sufficient in itself. Few philosophers and even fewer scientists would subscribe to a theory for psychological or metaphysical reasons unless the theory worked and worked well, and this the RF theory certainly did.

The use of frequency theories of probability made it possible for the first time to deal with cases like biased dice, probabilities of death, and statistical dispersions of stellar magnitudes. In short, the RF theory greatly expanded the number and types of cases with which probability theory could deal. (It might not be carping to point out that adherents of the new view generally continued to use the old methods, consciously or not, in cases where those methods had proven themselves.) As it happens, this expansion tended to shift the emphasis away from the traditional games of chance and towards the problems of actuarial and scientific statistics (I do not mean to imply that this shift is accidental). Thus the successes of the RF method tended to reinforce its theoretical claim to be 'scientific' and its psychological appeal to the admirers of science. No wonder it was embraced by so many scientists and mathematicians, and continues to appeal to many.

V

THE SUBJECTIVISTIC THEORY OF PROBABILITY

The subjectivistic theory of probability (SUB) identifies probability as the actual degree of belief in a given proposition held by some real individual at some specific time. Contrary to the usual pattern of scientific intellectual progress, the subjectivistic theory of probability has generally been developed *after* the objectivistic.

It is true that certain remarks about 'degrees of confidence' found in such early pioneers as Bernoulli and Laplace have led some researchers to catalog them as subjectivistic, or to criticize them as psychologistic',[1] but I have argued above[2] that this misconstrues their basically objectivistic classical theory. Instead, the first subjectivistic theory is generally held to be Frank P. Ramsey's essay 'Truth and probability'[3] (1931), which is self-consciously and intentionally subjectivistic in its insistence that probability measures the actual degree of belief of an individual. Ramsey also deserves the credit for showing how these degrees of belief can be compared and measured by studying betting behavior. Unfortunately, Ramsey's early death kept him from systematically elaborating his theory (or abandoning it, as one brief note suggested he might).[4] The principal developer, defender, and disciple of subjectivistic probability theory has, therefore, been Bruno de Finetti. De Finetti's friend and ally in the English-speaking world has been Leonard J. Savage. Together they made known (and somewhat respectable) the idea that there is no such thing as objective probability, only degrees of belief.

The subjectivistic theory has also been used in practical applications. The area which we call 'decision theory' relies on probability considerations in many cases. For some of these, it is sufficient to postulate that the agent has some ideally adequate knowledge of objective probabilities. But in actual experiments some researchers (such as Davidson, Suppes and Siegel)[5] have found it necessary or advantageous to consider actual degrees of belief, and to attempt to measure them.

In statistics, subjectivistic theories of probability have sanctioned widespread use of Bayes's Theorem (for reasons to be explained later) and have thus helped give rise to what is somewhat misleadingly called 'Bayesian statistics.' This school has grown steadily in the last few decades and it now seems that subjectivistic probability, though it is not sweeping away other contenders, is well established as a legitimate concept with at least some significant applications.

1 DEFINITION OF PROBABILITY

We have mentioned already the subjectivistic definition of probability as the degree of belief of a given person at a given time. This is the fundamental idea which distinguishes it from all other theories: probability is not objective, but depends essentially on someone's beliefs. Subjectivist theorists therefore deny that other theories of probability refer to anything at all. In de Finetti's emphatic slogan:[6]

PROBABILITY DOES NOT EXIST.

Usually the basic idea is restricted, so that not just any degree of belief can be counted as a probability. Additional requirements are:

1 Measurement – the term 'degree of belief' is only fully meaningful when one has specified how it is to be measured or obtained.
2 Coherence – Not just any degree of belief is admissible. Each person's set of degrees of belief in various propositions must agree with or conform to each other in a certain way.

The first point is stressed by both Ramsey and de Finetti. Ramsey argues that probability cannot be simply identified with 'the felt intensity of a belief' because (1) it is too difficult to quantify the intensity of feelings, and (2) strong feelings are not always associated with strong beliefs, as when we take things for granted.[7]

It is thus necessary to abandon felt intensity in favor of some other property of the belief. Both Ramsey and de Finetti favor a behavioristic approach in which the degree of belief is identified with the person's willingness to act on the proposition. This willingness to act can then be measured by observing behavior, especially in betting situations, which must be of a specific type in order to guarantee uniformity and comparability, and to avoid such complications as the diminishing marginal utility of money. The finished concept is thus 'operational', according to de Finetti,[8] and Ramsey says it resembles Einstein's concept of a time interval in that its precise meaning depends on how we specify that it is to be measured.[9]

The second restriction, 'coherence,' is the normative or regulative part of the theory. It is not enough to have different degrees of belief in P and not-P, we are told, we must also see to it that these degrees of belief do not conflict in a certain way.

Suppose, for example, John's belief that Gluefoot will win has the degree 3/4 and his belief that Gluefoot will lose has the degree 2/3. No formal contradiction exists, since no logically incompatible propositions are involved. Neither is it a psychological impossibility (as some have claimed it is psychologically impossible to believe P and not-P simultaneously, e.g.) as we shall now demonstrate.

If we adopt the betting method of determining degrees of belief, John's belief of 3/4 means that he will bet 3 of his dollars against one of mine that Gluefoot will win. But his belief of 2/3 requires that he also bet 2 of his dollars against one of mine that Gluefoot will *not* win. Most of us would not make such a pair of bets, of course, because they guarantee that John will lose money no matter what the outcome of the race. If Gluefoot wins, John wins a dollar on the first bet and loses 2 on the second, while in the other case he loses 3 on the first bet and wins only one on the second. In such a situation, gamblers would say I have made a 'Dutch Book' against John, a situation much to be deplored. But deplore it as we will, people do sometimes allow Dutch Books to be made against them. Usually it is because they are inattentive or confused, or do not understand the numbers involved, but sometimes they are just 'stupid' or 'irrational.'

The cardinal rule of subjectivistic probability is that one should not be irrational in just this way. The surprising result (which I. J. Good calls the Dutch Book Theorem) is that conforming to the probability calculus is a necessary and sufficient condition of avoiding the possibility of a Dutch Book. In the example just cited, the difficulty is that John's degrees of belief do not add up to one.

This violates the fundamental principle which de Finetti calls the theorem of *total probabilites*: 'in a complete class of incompatible events, the sum of the probabilities must be equal to 1.'[10] This theorem can be deduced from the requirement that one avoid the possibility of a Dutch Book, or, as the subjectivists put it, that one's beliefs be *coherent*.[11]

It is important to note that subjectivists do not think everyone's beliefs *are* coherent in this sense. What they are saying is that anyone who wishes to be 'consistent' or 'rational' in a certain heuristic sense must have coherent beliefs and will *ipso facto* conform to the probability calculus. Beyond this, each person is free to have whatever degrees of belief he or she chooses, and for any given set of propositions, an infinity of coherent probability distributions are possible. The freedom to believe what you will, subject only to the broad restraints of rationality, is a key feature of subjectivistic probability.

2 SOURCES OF INITIAL PROBABILITIES

Since a probability is nothing more than the degree of belief of a given person in a proposition, it exists only when someone has a belief, and it is established by determining what the degree of that belief is. Generally, subjectivists have tended to act as if each of us has an opinion about everything, because there are always some odds at which we will make a bet on any contingent proposition or event.[12]

When we ask for the source of initial probabilities we might intend either of these questions:

1 How do we establish a value for the degree of belief S has in P?

2 What is the source, cause, or explanation of the degree of belief which S has in P?

The answer usually given to the first is 'By a carefully conducted series of small comparative wagers.'

The answer to the second is 'It might be anything, but there are regularities.'

Now let us examine these in more detail.

The Subjectivistic Theory of Probability

In Dice Games

The probability for S that P (a Three will show on the die) is defined as S's degree of belief that P. If we wish to know what this value is, we *could* just ask S. But since people sometimes lie or suffer from selfdelusion, or may just be poor estimators of probability, a more reliable and 'scientific' method is to observe S's behavior – especially what he bets on P and what odds he requires. Experimental psychologists and decision theorists have developed methods for establishing personal probabilities, which are of considerable interest.

When the experimenters used a four-sided die, for instance, they found some subjects had a strong tendency to require odds greater than 3 to 1, though the 'objective probability' was clearly intended to be 1/4.[13] Also, the subjects exhibited behavior which seemed to be based on the old fallacy called the Maturity of Chance. Whenever a side appeared several times in succession, the subjective probability of that side 'would temporarily decrease for most subjects.'[14] (This attitude is particularly intriguing, since most theories claim that no adjustment should be made, and some claim we should *increase* the probabilities.)

By using these methods – setting up a series of bets and comparing them – we can establish to what extent S is willing to act on P. Or we can ask him. Or we can lead him to the crap table and see how he bets. In these ways we can establish what the initial probability is.

But *why* is the initial probability what it is? The second sense of our question about the source of initial probabilities is, 'What gives rise to, or accounts for, the probabilities we establish?'

Most of us think the probability of throwing a Three is 1/6. For many of us, the source of this belief is the voice of authority (Uncle Fred, the black sheep of the family) or just general enculturation. However, some have actually thought it through or worked it out for ourselves. In these cases, de Finetti says, the important thing is the judgment that all faces are equally probable. If this judgment is made, and if we are to have coherent beliefs (as he says we must) it is clear that 6 equal probabilities which sum to 1 must each equal 1/6.[15] As to *why* we judge things to be equally probable, we can only say that we have a general tendency to do so and that considerations of symmetry weigh heavily in particular cases. Furthermore, we have a great store of experience with 'fair' dice, and we know that they are generally manufactured with the express aim of making the sides equally probable.

Given all this, it is little wonder that we tend to assign the value 1/6 to the probability of throwing a Three. But subjectivistic probability theories do not tie us to that value. They share with the relative frequency (RF) theories the ability to deal with biased or irregular dice. If a die persists in throwing a large number of Threes we are free to raise our subjective probability of *P*. The subjectivist thus has an advantage over the Classical theorist in a crooked dice game, because he is free to adjust his probabilities. But there are and can be no *rules* telling the subjectivist when or even in what direction to adjust his probabilities. The same freedom (license?) which enables him to adapt to the crooked dice game also legitimizes the fallacy of the Maturity of Chances and any superstition or irrational preference the individual might have. In a subjective assessment of probabilities, any factor is relevant which the individual takes to be so, and any (coherent) value is just as good as any other.

In Actuarial Cases

Even in the most abstruse cases where computers and statisticians labor collectively over arcane and complex formulas, we must remember that in the end a probability is just one person's opinion. The complicated statistical laws that are taught in universities are, according to Savage, just elaborate reformulations of the coherency reguirements.[16] The amassing of huge bodies of data and the compulsive plotting of graphs and frequencies can be accounted for, according to de Finetti, by the simple psychological fact that we generally expect frequencies to continue,[17] or to put words in his mouth, we count and measure the present because we expect the future to resemble the past.

Most non-statisticians would be surprised to learn how many statistical issues and decisions depend on taste rather than logic and are decided by such psychological considerations as simplicity, ease of computation, appeals to authority, and intuitive approval rather than by rigorous mathematical proof.

As a true American pragmatist, I would also like to point out how much of statistics is justified and accepted simply because it works. The businessman evaluating production and the scientist estimating probable errors may be swayed by the simplicity of a technique but, in the end, they will use the methods which best serve their purposes, regardless of mathematical pedigrees. Nearly all of applied statistics – and other forms of applied mathematics – was developed for practical reasons. We are interested in it because it works. Its working may be grounded in a deep metaphysical correspondence between mathematics and the world, or it may just be a lucky coincidence. In either case we would continue to study and develop applied statistics because of its great practical importance. The same is true of statistical probability (as de Finetti should have said, but didn't). We accept certain ways of computing probabilities just because we like them – but we like them because they work. The selection procedure is indeed based on our desires and feelings, but that doesn't mean it is irrational.

Subjectivistic theories can account for actuarial probability by these two arguments: much of it is just mathematical elaboration of the coherency requirement; the rest is just a matter of taste. But beyond this, the subjectivist viewpoint actually expands the range of statistical probability, by allowing much greater use of Bayes's Theorem.

Some objectivists have called for a greatly restricted employment of Bayes's Theorem because the H_j factors, which are called initial probabilities, in the formula are generally not known, If we have no measured frequencies (in the RF view) or computed values (in the *a priori* and Classical views) for these we have no values to plug into the formula and thus should expect no answer.

But on de Finetti's view there are no objective probabilities at all; hence, *a fortiori*, there are no unknown objective probabilities. If we wish to know the value of H_5, for example we just ask ourselves what it is. (If we wish to be careful, we could conduct a series of comparative wagers.) Once we establish a (subjective) value for H_1 through H_j, we are free to compute the desired probability.

Objectivists recoil in horror from this procedure, since the output will be 'tainted' by the subjectivity of the input. But for the subjectivist, *all* probabilities are subjective so that is certainly no reason for rejecting Bayesian methods. Furthermore, once the feeling of repugnance is overcome, the subjectivist thinks he can offer the

225

objectivist good reasons for *accepting* the procedure. In most cases which involve a good deal of data or empirical evidence, it can be shown that the initial subjective values of the prior probabilities are 'swamped' by the accumulation of objective data and play rapidly declining roles in determining the answer.[18] The solution, then, can be made as 'objective' as one pleases – all you have to use the subjective values for is to 'get started.'

Bayesian statistics is a growing field, whose techniques are applied in many areas. Of the polemics raised against it, few have argued that it does not work, or that it in fact gives the wrong answer most of the time. No, the objection to the Bayesian method is that it is not *justified*, since it doesn't conform to the particular theory of probability (usually an RF theory) held by the critic. But the Bayesian method *is* justified in the framework of subjectivistic probability. Because of this the concepts have become so intermingled that for many scholars 'Bayesian' and 'subjectivistic' have come to be synonymous. The reader should recall, however, that objectivist theories do sanction the application of Bayes's Theorem in those cases where the initial probabilities are known, so not all Bayesian solutions are subjectivistic. Otherwise, the terminological confusion is of no great importance.

There is another way in which the subjectivistic theory claims to have an advantage over objectivistic statistics. You may recall from our discussion of RF theories the problem of the reference class: If 10 per cent of American Indians and 30 per cent of physicians drink Scotch, what is the probability that an Indian physician does? We concluded at the time that rules of thumb might help, but there can be no general principle establishing which is the 'correct' reference class or how the two should be merged. Now de Finetti and company claim that the problem of the reference class is solved (or at least dissolved) by the simple recognition that selection of the reference set is *always* arbitrary.[19] The objectivist found it puzzling because he was blinded by the 'prejudice' that there must be some true or objective value of the probability, if only he could figure out how to compute it. The subjectivist, however, recognizes that his opinion is the final authority, and is free to consider or ignore any data about any classes whatsoever. There is no *correct* reference class since there is no correct probability. This is so even for those highly uniform occurrences which objectivists call 'repetitive events' – but that's a problem for subsequent sections.

3   PROBABILITY OF SINGLE EVENTS

SUB shares with the a priori theory the advantage that it permits us to speak of the probability of any event whatsoever, simple or compound, unique or repetitive. If someone has an opinion on the matter, that opinion has a degree of belief, and the event therefore has a probability.

That Smith will be elected Mayor

It is on this example that the subjectivistic theory of probability really shines. The Classical and RF theories lead to the embarrassing conclusion that there *is no* probability that Smith will be elected. Anyone who reads the papers in an election year knows that this supposedly non-existent value is not only treated by most people as real, it is the subject of a great deal of discussion, controversy, and difference of opinion. The a priori theory does somewhat better by admitting that the probability exists, but it too abandons the field when it comes to setting up a value. The enormous complexity of the situation and of ordinary language make calculation impossible.

SUB seems closest to common sense in allowing straightforward talk about the probability that Smith will win.[20] After all, any event we can have an opinion about is an event we can attribute probabilities to in ordinary discourse.

The difficulty, however, is that SUB does not allow for just one value of the probability – there can be as many probabilities as there are opinions. The editorialists who think they are in disagreement about Smith's chances really are not at all. There is no such thing as 'Smith's chances' to disagree *about*. There is the *Star*'s opinion of Smith's chances and the *Journal*'s opinion of Smith's chances, but like my taste for apples and yours for oranges, these are not contradictory opinions about a matter of fact, but non-disputable differences of taste.

There is an obvious parallel here between subjectivistic theories of probability and emotivist (or non-cognitivist) theories of ethics. If we take Stevenson[21] as the spokesman for the emotivists and consider a parallel dispute about whether Smith is (will be) a 'good' mayor, we have the following situation:

1 There is no 'matter of fact' about whether or not Smith is a good mayor.

2 When I say 'Smith is a good mayor' I am expressing my emotive approval of Smith.

3 When you say 'Smith is a bad mayor' you are expressing your emotive disapproval of Smith.

4 You could win the argument by showing that my views are logically inconsistent – no other argument is decisive.

5 You *might* win the argument by showing I am mistaken about the facts – some disagreements in attitude are rooted in disagreements in belief.

6 It is possible that we might agree on all the facts and still legitimately disagree on Smith's moral worth.

Now if we quickly make the appropriate changes, we have a subjectivistic discussion of the probability that Smith will be elected mayor.

1 There is no 'matter of fact' about the probability that Smith will be elected.

2 When I say 'Smith's probability of election is 0.6' I am expressing my degree of belief that Smith will be elected.

3 When you say 'Smith's probability of election is 0.3' you are expressing your degree of belief that Smith will be elected.

4 You could win the argument by showing that my views are incoherent – no other argument is decisive.

5 You *might* win the argument by showing I am mistaken about the facts – some differences in degrees of belief are rooted in differences of belief about the facts.

6 It is possible that we might agree on all the facts and still legitimately disagree on the probability that Smith will be elected.

The striking similarity between these theories is of course due to the fact that each takes a concept which is normally held to be objectively present in the world and reinterprets it as a subjective attitude of the mind. Not surprisingly, both are vulnerable to the general criticisms that can be levelled at subjectivism, as we shall see in the penultimate section.

That Smith will roll a Five

Assessments of this probability generally agree, which accounts for the popularity of books on probability theory. But they sometimes disagree, which accounts for the popularity of crap games. What the subjectivistic theory cannot do is tell us how to arrive at the correct assessment.

Consider someone who is forced to wager on whether or not Smith will roll a Five. De Finetti has said[22] he doesn't intend to explain why it will or will not be a Five, but he also has said nothing that will guide him in his choice! The only normative advice the subjectivist offers is the injunction that our beliefs should be coherent. It is admirable and remarkable that this one bit of guidance forces us to accept and obey the calculus of probabilites. But remember, I have argued that the calculus is of little interest to us because it tells us nothing about what probabilities are or how to measure them. So far, no help.

Now subjectivism does try to explain what probabilities are: they are the degrees of belief that real persons have in propositions. Suppose, then, we tell our bettor that the probability that Smith will roll a Five is equal to (identical to) the degree to which he (the bettor) believes Smith will roll a Five. We have cast him back upon his own resources, then, with no advice except to do what he thinks best.

The Classical theory told him *how* to judge his chances: divide the number of favorable possibilities by the total number of possible outcomes to get a probability of 1/6. This advice is very helpful and successful until someone slips in a loaded die.

The RF theory can deal even with this latter contingency: count the frequency of successes in the past and project that relative frequency as the probability of Fives in the future. That's good when a lot of evidence is available, useless on the first roll. Still, it's better than nothing, and nothing is what you get from the subjectivists.

Let me say it again, since it slips past most people who fasten on other debatable or deplorable aspects of subjectivism: SUB gives no advice at all on how any initial probabilities should be obtained! All it does is argue that each of us should obey the probability calculus. But if our bettor should repeatedly bet large amounts of money that Smith would throw a Five and give the explanation that he was sure he would win because the probability of a Five is 0.9 and if years of this losing enterprise laid waste his large fortune and left his family destitute, then we still could not accuse him of acting unwisely, After all, if acting on probabilities is ever rational, it is surely rational for a wealthy man to bet on a probability of 0.9. And according to the subjectivistic theory of probability, the probability was always 0.9 *because he said it was.*

I can see already the pained expression on the faces of the subjectivists who want to explain to me about long series, Bernoulli trials, and Bayes's Theorem. The bettor *would* be irrational, they say, since his belief that the probability was 0.9 is incoherent with the universal belief that events with a probability of 0.9 don't persist in failing 84 per cent of the time, and a careful attention to the results would convince him that he should alter his view of the probability of Fives.

My first response to this depends on a point I have made before; it is *not impossible* for repeated trials of a 0.9 probable event to result in 84 per cent failures, it is just very improbable. The Law of Large Numbers tells us that the odds are 9999...to 1 against it, for long runs, so that it would take incredible bad luck for it to happen. But of course our bettor believes he is having incredible bad luck!! His beliefs are, therefore, as consistent and coherent as anyone's.

My second response is that consideration of past failures might convince him that the probability in the past had been around 0.16. Even so, it would be neither inconsistent nor incoherent for him to say each time that the probability of the *next* throw being a Five is 0.9, since SUB gives no rule for connecting past experience to present probabilities, and allows him to continue using any value he chooses. In fact, de Finetti's theory might encourage him to act this way, since he says explicitly that there are no such things as repetitive events, a point we will now discuss.

4 PROBABILITY OF REPETITIVE KINDS OF EVENTS

As a matter of terminology and principle, de Finetti refuses to talk about repetitive events. A probability is the degree of our belief in an event (proposition) and that belief is not tied down by any similarity properties, class membership, or other objective features of the event. If we think the probability of Heads on this toss is 1/2, it is still perfectly all right to think that the probability on the next toss is 1/3. There need not be any fixed probability, because each event is an individual and must be judged as such – not as part of a set.

As de Finetti puts it, statistical events are never 'identical', they are at best 'analogous.'[23] A more picturesque expression of the same sentiment was made by the folk humorist Brother Dave Gardner,[24] who complained about 'folks who say "Let's do this agin sumtime"' on the grounds that

> 'You cain't do that.'
> 'What?'
> 'Again.'
> '!??!'

'You can do somthin simular, but you cain't do that again. Once it's gone, it's gone.'

Nevertheless, it is obvious that people do treat certain classes of events collectively, and de Finetti tries to explain how this works in his system.

That a Thirty-Year-Old will get Married

The purest subjective judgment is our estimation of the probability that *this* thirty-year-old, say Jack Warndof, will get married. It is obvious to everyone who knows Jack that he is a special case, a unique individual – indeed, an oddball. The probability that he will get married depends on his peculiar circumstances and his unusual personality and has little to do with statistics from the Bureau of the Census.

This example shows the appeal of SUB in dealing with individual cases, a virtue we discussed in an earlier section. But, for all his individuality, Jack really is one of the thirty-year-olds who make up the classes, frequencies and percentages of the census bureau. What can we say about this collective aspect of his existence?

First of all, it is arbitrary *which* collective we concern ourselves with. We may wonder about any of the groups to which Jack belongs or we may, as the title of this section suggests, treat only his age as relevant. (De Finetti repeatedly stressed that objective probabilities are impossible in part because judgments of relevance are ineliminably subjective.) But once we have selected the reference class, there is a natural tendency to assume that the frequency of marriage for that group will continue to be fairly stable. This remains an unexplained psychological fact:

A rich enough experience leads us always to consider as

231

probable future frequencies or distributions close to those which have been observed.[25]

Finally, we must remember that even a firm faith that 12 per cent of thirty-year-olds will continue to marry each year does not commit us to believing that the probability that Jack or any other thirty-year-old will marry is 0.12. There is no formal, theoretical, connection between frequencies and probabilites at all in this system. Frequencies may influence our subjective assessment, but they need not determine it.

That a Dice Throw will be Five

Now that I have stressed so heavily the subjectivists' view that there are no repetitive events, the reader is no doubt wondering if we must give up Bernoulli's Theorem, the Laws of Large Numbers, and other formulas which depend on constant probabilities. The answer is that although we are not *required* to treat any series of events as equiprobable, we are *free* to do so if we choose.

Rather than talk about equiprobable or Bernoullian events, de Finetti introduces the term 'exchangeability' and defines it as follows:[26]

A collection of events is said to be exchangeable *if the probability Wh that h of them occur, depends only on h and not on the particular events chosen.*

In all those cases where objectivists insist upon the equiprobability of repetitive events, subjectivists are free to judge a set of events to be exchangeable. Obviously, exchangeable events will obey the same logic and mathematics as those repetitive events earlier banished from the fold. The advantage of the new terminology is that it permits us to describe the situation without introducing constant, objective probabilities. If a person judges a set of events to be exchangeable and of a certain subjective probability, then the coherence requirement forces him to make the same predictions about groups and series of occurrences as objectivists would make using Bernoulli's Theorem.[27]

As it happens, most of us do judge events like dice throws and draws of a card to be exchangeable, because of our long shared experience and because of considerations of symmetry. We are therefore in a position to use calculations of this sort quite frequently. The subjectivist difference is that no one is ever *forced* to accept these calculations, since we are always free to vary our attitude about any or all or the events. The fundamental reality remains the individual's opinion about an individual event.

5 ABSOLUTE PROBABILITY AND PHYSICAL CHANCE

All is subjective – nothing is absolute.

In a system which defines probability as the individual's degree of belief in a proposition, it is obvious that there can be no one answer to 'what is the probability of X?' There are as many answers as there are beliefs, and no answer is better than any other (coherent) answer, since the individual is theoretically free to hold any opinion whatsoever.

One early writer, Emile Borel, who is sometimes classified as a subjectivist, did try to give some sense to 'objective probability' by saying that it referred to 'the probability which is common to the judgments of all the best informed persons, that is to say, the persons possessing all the information that it is humanly possible to possess at the time of the [judgment].'[28] These probabilities are like physical constants in that scientists agree on their present value but may change them in the future light of 'the progress of physical-chemical theory.'[29]

Since Borel was writing before Ramsey and de Finetti, and is less than explicit and consistent in his subjectivism, I think we can safely dismiss his view as non-standard. (Alternatively, we could say that anyone who has *that* view of physical constants doesn't understand the word 'objective' anyway.)

Savage and de Finetti are more definite and consistent in their views and, for them, objective probabilities are an illusion, a superstition.[30] De Finetti, indeed, delights in pointing out that even in 'objectivistic' systems there is no such thing as '*the* probability of X,' since the choice of a reference class is always arbitrary and the amount of evidence is always insufficient. Furthermore, if 'objective' is taken to mean 'capable of being judged true or false on the basis of a well-determined observation which is at least conceptually possible,'[31] then statements of subjective probability are even more objective than their RF counterparts, since one *can* gather sufficient evidence to confirm or refute them.

Of course a completely confirmed objectively true statement about a subjective probability would deal with an individual's degree of belief in X, not with X itself. It would be 'the probability of X for S' not 'the absolute probability of X.' For this latter concept there is no place at all in the subjectivistic scheme of things.

As to physical chance, de Finetti, at least, is unwilling to pass judgment. It is not just that we are unsure about the answer to the determinism/indeterminism dispute, but, more importantly, he sees the question as irrelevant to probability theory,

> because, whatever the explanation of the uncertainty might be..., the sole concrete fact which is beyond dispute is that someone...feels himself in a state of uncertainty, and has to decide on and adopt some point of view as a basis for previsions and related decisions.[32]

The nature of the physical universe is a question for cosmologists and metaphysicians – subjectivistic probability theory will work in either case.

6 THE METAPHYSICAL STATUS OF *P*

The basic fact is that P is an assertion about someone's mind, not about the event or proposition which is the formal subject of P. If I say 'Gluefoot's probability of winning is 0.6' the subjectivists agree that the content of this sentence is the assertion (perhaps the expression) of my own degree of belief in Gluefoot's winning. Since, as we would normally say, my belief could be well- or ill-founded, in agreement or disagreement with the facts, etc., it is clear that an expression or assertion of that belief has no necessary connection to the facts. It is not 'objective' in the root sense of being about or dependent upon the objects in the world, but is rather subjective in the sense of being about or dependent upon the psychological subject.

Obviously the *exact* metaphysical status of P depends on one's views about the mind/body problem. If some version of the identity theory is true, then these statements, which seem to be about horses, are really about beliefs, which in turn are really certain states of the central nervous system (or whatever). Thus the ultimate reality which determines P's truth is physical. Alternatively, if some version of dualism or idealism is true, my statement about Gluefoot depends for its truth upon the existence of a certain *mental* state of affairs, namely my belief of the appropriate degree.

This metaphysical adaptability is captured by de Finetti's assertion that a probability is nothing but an opinion,[33] and thus has whatever ontological status an opinion does. But it is put even more forcefully and graphically by I. J. Good who argues that we need not get involved in 'metaphysical problems concerning mind' because subjective probability can occur in any 'communication system that has apparently purposive behavior,' including 'Martians or machines.'[34]

The essential feature is that probabilities are not tied to external reality, but depend entirely on whatever internal realities one sanctions in one's metaphysics.

7   THE EPISTEMOLOGICAL STATUS OF *P*

In assessing the epistemological status of *P*, we need to distinguish two main concerns:

1 How do *we* know *P*? That is, how can observers establish the degree of *S*'s belief in *X*?

2 On what is *P* founded? That is, what steps, procedures, and processes go into forming *S*'s belief and can they be comparatively evaluated?

Addressing the first of these concerns, the simplest way to establish what *S* believes is just to ask *S*. If *S* is honest, helpful, and good at introspection, he might be able to tell us precisely what degree of belief he has in *X*, and we would be justified in believing him.

Unfortunately, we are not all good introspectors. As Freud and others have shown us, we are seldom aware of all our motives, attitudes, and beliefs. And even if *S* knows the degree of his own belief, many motives, fears, or incompetencies might prevent his conveying that knowledge to us.

For all these reasons, most subjectivists have agreed that the best way to establish *P* is not to ask *S*, but to observe his behavior.

Since the time of Ramsey, the usually advocated method has been the study of betting behavior. There are many advantages to this method, but perhaps the most important are

(a) It is in the subject's interest to act on his own best judgment, thus reducing deception and inattention;

(b) It is possible to manipulate and compare the bets in such a way that the degree of belief (if stable) is revealed with considerable quantitative precision.

Experimental psychologists and others have devoted a good deal of time and energy to establishing procedures which are useful and precise and reduce the influence of such factors as the diminishing marginal utility of money and a like or dislike for gambling as such. These studies have shown that most people do have fairly stable degrees of belief which can be ascertained without too much trouble – at least in a laboratory situation.

Turning now to our other problem, how does S form his opinion? What factors shape his belief, and how can the process be evaluated? There are two levels to the answer to these questions.

First of all, the theory itself sets no limits and imposes no rules on anyone's opinions, save only that we must maintain coherence. Otherwise, we are in principle free to believe any proposition to any degree whatsoever. The theory is epistemologically neutral, or devoid of content. Like formal logic, it tells us that if we believe certain things, we must believe certain others, but the choice of premises is always up to us (for reasons such as this, Ramsey has joined many non-subjectivists in suggesting that probability theory should be 'taken as a branch of logic, the logic of partial belief and inconclusive argument'[35]).

Although the theory itself says nothing about the source and justification of our beliefs, that does not mean there is nothing to be said. De Finetti, in a descriptive vein, talks about various factors which shape our opinions. Shared experience, constant frequencies, common character traits, a concern for symmetry and an acceptance of the opinions of others all conspire to account for that vast body of shared opinion about probabilities. What others thought betokened an objective regularity, de Finetti reinterprets in terms of psychological regularity.

Finally, as to the normative question, what can we say about the merits of different probability appraisals? Clearly the theoretical answer is 'nothing'. One probability is just as good as another in the eyes of the theory, provided both are coherent. But obviously one probability is *not* just as good as another. Some bets, plans, decisions are better than others. In a game like 'Go' the rules leave us free to use any strategy we like, but some strategies are more productive than others. In the game of life, likewise, the rules of the probability calculus leave us free to make any probability assessment we choose, but some are reasonable, some not. De Finetti makes some attempt to indicate the kinds of things we should consider in judging probabilities,[36] but always refuses to allow any general rule of assessment. Ramsey, on the other hand, is willing to say that a judgment is praiseworthy if it comes from a 'good habit.'

Thus given a single opinion, we can only praise or blame it on the ground of truth or falsity: given a habit of certain form, we can praise or blame it accordingly as the degree of belief it produces is near or far from the actual proportion in which the habit leads to truth. We can then praise or blame opinions derivatively from our praise or blame of the habits that produce them.[37]

The final answer, then, seems to be the pragmatic one that an appraisal is good if it leads to success in practice, and a method for picking winners is good if it picks winners. Since there are no objective probabilities to be conformed to, our judgment must be based on practical results rather than correspondence to reality. The epistemology of *P* runs rather more towards the Pragmatic than the Correspondence Theory of Truth.

8 THE RATIONALITY OF PROBABILITY BEHAVIOR

Experience teaches us that in many situations individuals must act under conditions of uncertainty about the present, the future, and the consequences of their actions. Much of the theory of probability is aimed at explaining, formalizing, rationalizing such behavior. The RF and AP theories tell us where to look for information, how to calculate probabilities, and what conclusions we must reach. The subjective theory is less overbearing: it doesn't tell one what to do at all, it only tells one what one *must not* do, namely violate the rules of coherence which make up the calculus of probability.

Although it is thus clearly normative, subjectivistic probability theory does not seek to be authoritarian or moralistic.

The 'one must' is to be understood as 'one must if one wishes to avoid these particular objective consequences.' It is not to be taken as an obligation that someone means to impose from the outside, nor as an assertion that our evaluations are always automatically coherent.[38]

The force of the coherency requirement springs from the logic of rational choice. If 'choice' and 'desirable' are to make any sense at all, it must not be rational to choose the least desirable alternative.[39] But if one accepts a Dutch Book, one is choosing a certain loss (losses are by definition undesirable) and is therefore acting irrationally.

The point of coherency requirements is to keep us from acting the fool by thwarting our own desires. We remain free to desire what we will, and to estimate probabilities as we choose. All we must do is be consistent in our desires and expectations.

Subjectivistic probability theory is, like deductive logic, a minimal standard of rationality.

9 CHIEF CRITICISMS OF SUBJECTIVISTIC THEORIES

The most basic criticism of subjectivistic probability is that it confuses feeling with fact.

It is both true and important that people have different opinions about Gluefoot's chance of winning (that's what makes a horse race). Experimental psychologists have done much good work in starting to sort out the probability and utility functions people 'act as if' they have, and the evidence supports what professional gamblers could have told us all along: people have different ideas about the odds, most of them pretty good, some downright stupid. The particular kind of stupidity which consists in accepting a 'Dutch Book' or 'sure loss' gladdens the heart of the grifter and the conman – but it's not what supports the gambler! Few of us are so soft a mark that we accept certain losses. But many of us believe in the maturity of odds, betting a martingale, or drawing to an inside straight. The gambler doesn't cheat or deceive us, and we need not be incoherent fools to lose our money – the gambler understands the odds and plays the percentage and, in the end, will always take our money *because the odds are real.*

If your sainted mother suddenly started giving 4 to 1 odds on the toss of a fair coin, you would be well advised to protect your inheritance by urging her to stop, because she will certainly lose in the long run. But according to the subjectivistic theory of probability there are no grounds for criticism at all in this case. If she thinks the probability of heads is 0.2, then the probability of heads *is* 0.2

and there's an end on it. There are no objective probabilities to invoke; there are no facts to be considered; there is no counsel to be gained from probability theory.

Surely this is mistaken. The poor woman will lose her money, not because I think so or you think so, but because it really is a fact that a fair coin will almost certainly show Heads in more than 20 per cent of a long run of tosses. Emphasis on the psychological fact of diverse opinions blinkers the ordinary fact that some things *really are* more probable than others. To flout this ordinary fact is not the same kind of irrationality as having incoherent opinions or believing *P* and not-*P*, but it is irrational none the less, just as it is irrational to ignore any obvious fact about the world.

It might be said in defense of the subjectivists that they don't deny the irrationality of Mama's behavior, they just deny that probability theory as such can correct it. But what discipline *can* help us assess likelihoods and probable outcomes, if not probability theory? When subjectivistic theorists refuse to tell us how to assess initial probabilities they are 'copping out,' in the current slang. They are leaving us with only the calculus of probability as a minimal guide to rationality. Now the probability calculus is very important – indeed it is common to all theories of probability. But that very universality means that it cannot be used to individuate theories. As I have repeatedly argued, the true test of a theory is how it defines 'probability' and how it establishes initial probabilities. Subjectivistic probability theory tells us that probabilities are degrees of belief and are established by psychological investigations.

This is a psychologism.

A psychologism is the mistaking of a belief or opinion for a fact or objective condition, or the other way round. In logic, it is psychologistic to talk about beliefs and laws of thought rather than objective deductive relationships. In probability theory, it is psychologistic to confuse people's beliefs and behavior with the objective grounds of those beliefs. It is false that the probability of Heads is what Mama thinks it is, and that is why she is losing her money. Her opinion is deplorable because it is wrong; but it could not be 'wrong' unless there were some 'right' to compare it to, and that must be objective probability.

If de Finetti were right in his assertion that objective probability does not exist, then we would indeed be free to believe anything we chose, with whatever degree of confidence we felt, because it would not matter what we believed. But it *does* matter what we believe (as Mama's incipient bankruptcy demonstrates), therefore de Finetti must be mistaken (*modus tollens*).

The probabilities which guide our lives are as 'real and earnest' as life itself. They are not matters of whim or opinion. The physician does not accept the patient's 'degree of belief' that the tumor is benign; the astronomer does not ask the janitor for the probability that a cepheid variable will go nova. These probabilities are 'objective' in a way which makes 'subjective' a pejorative. They are objective in a way which shows that subjective probabilities could never serve as a basis for science. Scientists must always assume that there is a 'fact of the matter' which scientific investigation pursues and which scientific opinions approximate more or less correctly according as they are better or worse opinions. But de Finetti's dictum would have it that there is no 'fact of the matter' in probability cases and therefore that all opinions are on an equal footing. This cuts at the very root of thought and removes the whole point of the scientific enterprise. No, subjectivistic probability will not do for science.

Finally, to mitigate the criticism somewhat, we should agree that it is sometimes valuable and important to enquire into what people think rather than what is the case. In deciding whether to vote for Betty Castor or Julian Lane, it is important to ask which *is* the better person or *will be* the better legislator. But in *predicting* the Castor–Lane outcome, it is at least as important to ask what people *think* about the qualifications and desirability of each.

In like manner, opinions about probability can be a proper object of study for experimental psychologists. The findings of such researchers as Davidson, Suppes and Siegel give us valuable information about what people *think* is the probable outcome of casting a four-sided die. But they do not (except incidentally) tell us what *is* the probable outcome of such a cast.[40] In learning such things we learn about people, not probabilities. The fundamental mistake of the subjectivists lies in confusing the two.

10   CHIEF VIRTUES OF SUBJECTIVISTIC THEORIES

Despite its fundamental misconception, subjectivistic probability theory has at least four virtues:

1 It accommodates talk of the probability of individual events.
2 It expands the field of possible applications of Bayes's Theorem.
3 It has made contributions to the growth of decision theory and psychological knowledge.
4 It shows clearly why we must act on the probability calculus to be rational.

First, consider the problem of unique events, which we have referred to throughout this book. Savage offers an example:[41]

> on no ordinary objectivistic view would it be meaningful, let alone true, to say that on the basis of the available evidence it is very improbable, though not impossible, that France will become a monarchy within the next decade... The personalistic view claims, however, to analyze such statements in terms of mathematical probability, and it considers them important in science and other human activities.

This is indeed a telling point against RF theories and others which reject isolated probabilities.[42] It is a sin against ordinary language to deny the meaningfulness of such assertions, and merely an evasion to say that they must involve some concept other than probability. Whether or not subjectivism succeeds in dealing with the problem, the reader must decide for himself, but surely it is better to try to explain it rather than sweeping it under the rug.

Second, subjectivism sanctions and justifies a type of statistical inference which has proven useful though controversial. Bayesian statisticians have given us many fruitful and accurate calculations. Their opponents have argued that whatever the results, the methods are theoretically unsound. Practicing statisticians, operational researchers, and cut-and-try mathematicians have gone ahead and done it anyway, without regard to theoretical niceties. Now many of these thinkers are sliding into subjectivism. Since the theory justifies a desirable method, the theory becomes desirable.

The success of Bayesian methods is understandable. Mathematically, it makes little difference what values we plug into Bayes's formula as initial probabilities. What is not easy to explain is the theoretical justification for pulling such figures out of a Bayesian bowler. If subjectivism continues to be the only theory which allows people to do what they really want to do anyhow, there will be a powerful incentive for its continued growth, despite its manifest disadvantages.

Third, the interest in SUB has been an added stimulus for research into decision theory and the like. As Carnap puts it, the subjectivistic theory is 'of importance for the theory of human behavior, hence for psychology, sociology, economics, etc.'[43] Evidently this value is not unique to SUB, since researchers who do not deny the existence of objective probability (such as Davidson, Suppes, and Siegel) engage in the very same kind of experiments as those who deny it (such as de Finetti). It would seem, then, that this virtue is more heuristic than theoretical.

Finally, it seems to me that no theory does a better job of explaining why we should accept the probability calculus. It is all very well for Kolmogorov to derive the calculus from clear and self-evident axioms, or for Reichenbach to show that it follows from the mathematics of infinite series, but I think that most of us gain a greater conviction when shown that otherwise a sharpie could turn us into certain losers. Many of us are hostile to mathematics, but no one wants to be a sucker!

VI

CONCLUSION

The purpose of this chapter is to present my own views on the nature
of probability's philosophical foundations, now that I have tried
earnestly to present the major alternative theories as clearly and
objectively as possible.

Let me begin with a few remarks about foundations, explications,
and reductions.

As Quine has noted, one who seeks the foundations of mathematics
engages in an effort which is dubious from the beginning, for 'Where
might he find foundations half so firm as what he wants to found?'[1]

I think it is manifestly evident – if a bit humbling – that no merely
theoretical or philosophical discourse could bring down and eliminate an established mathematical practice unless some alternative
means are suggested for doing what has been getting done. Hence,
we need have little fear of finding that probability theory rests on
shaky foundations, or even none at all; for the response of the
practicing mathematician will surely be a shrug of the shoulders and
continued use of probability theory. In fact, many mathematicians
and philosophers already take the view that probability neither has
nor needs any philosophical foundations – it is an uninterpreted part
of formal measure theory which we use whenever we find it
convenient, informative, practical. Such people get on very well in
probability theory, and have even written quite good textbooks,
despite their professed inability to explain what they are talking
about and their lack of interest in trying to find out.

Nevertheless, many will continue the search. There is in human
beings an endless well of curiosity, an urge to ask 'Why is this so?'

and 'What does that really mean?' When people of this bent encounter probability theory they always ask 'But what is probability theory and why does it work?' This book is an exposition and analysis of the major attempts to answer such questions. We have seen philosophers attack the problem in many ways, but in the end each has offered some definition, reduction, or explication of the concept of probability itself. This is obviously the key, for since the time of Kolmogorov it has been clear that all our theorems and calculations can be reduced to a set of basic axioms employing only one undefined notion, that of probability itself.

Our major problem, then, is the definition, reduction, or explication of the term 'probability.'

The art of defining terms can be divided broadly into two somewhat different activities. One can *find* a definition of the descriptive/lexicographic sort, or one can *give* a definition of the prescriptive/stipulative sort. The former activity is of genuine linguistic and historical interest, but it cannot be expected to generate important philosophical truths no matter what the shallower of the ordinary language philosophers would like to believe. A typical example of the latter would be the CTP's official definition of 'probability' as the ratio of all favorable to all equipossible cases. The notorious problem of giving 'equipossible' sufficient sense to make the definition both non-vacuous and non-circular is a good example of the difficulties attending any effort to reduce our confusion by giving definitions. Because of such difficulties, philosophers such as Keynes and Lewis have held that 'probability' is *sui generis* and undefinable. We *have* a notion of probability (whether linguistic or intuitive), but we cannot hope to break out of the circle of synonyms to explain that concept because it doesn't depend on anything else for its force and validity. Whether or not this is so, we certainly have been unable to agree on a definition of 'probability' which comes anywhere near to meeting our philosophical needs.

An alternative foundational maneuver is to seek to reduce the problematic theory or concept to some more basic theory or concept which we hope will be better understood. This notion of reduction gained great currency among the Logical Positivists and has been the object of considerable philosophical attention. Briefly, we may say that one theory is reduced to another when every concept of the original theory can be definitionally connected to some concept(s) of the reducing theory, and every law contained in the original theory can be replaced by an empirically identical law in the reducing theory. In the happiest situations, the reducing theory will be clear and unproblematic, but even if this is not the case, it is commonly held that gains in theoretical unity and ontological economy which can be achieved by this procedure are sufficiently valuable to justify the effort.

Clearly the best example of an attempted reduction in our subject is Reichenbach's effort to reduce the theory of probability to the theory of relative frequencies in infinite series. He contended that such a reduction enabled us to say in the clear language of mathematics whatever we had wanted to say in the fuzzy language of probability, and to solve the application problem by showing that probability theory is applicable if and only if there exists an infinite series leading to a limit. Unfortunately, this reduction severely truncates probability theory, by ruling out of court those efforts to apply probability to single cases or even to finite populations. It also has severe epistemological and metaphysical difficulties. Still, it remains the one attempted reduction which has met with considerable acceptance and continues to satisfy the foundational urges of many philosophers, mathematicians, and scientists.

An explication of a concept is a little more than a definition, a little less than a reduction. It seeks a definition indeed, but one which lies between the lexicographer's descriptive definition and the pure theorist's stipulative one. An explication should be faithful to standard usage, while at the same time making it more precise and understandable. As Nelson Goodman puts it[2]

> If we set out to define the term 'tree', we try to compose out of already understood words an expression that will apply to the familiar objects that standard usage calls trees, and that will not apply to objects that standard usage refuses to call trees. A proposal that plainly violates either condition is rejected; while a definition that meets these tests may be adopted and used to decide cases that are not already settled by actual usage....
> [There is a] characteristic dual adjustment between definition and usage, whereby the usage informs the definition, which in turn guides extension of the usage.

Rudolf Carnap sees himself as engaged in explication of this kind:[3]

> By an explication we understand the transformation of an inexact prescientific concept, the explicandum, into an exact

concept, the explicatum. The explicatum must fulfill the requirements of similarity to the explicandum, exactness, fruitfulness, and simplicity.

If a foundational effort were to succeed in achieving an explication of this sort, it would provide us with a clearer conceptualization of probability, which would encompass most existing usages and guide our future determinations.

Do any of our suggested theories count as adequate definitions, reductions, or explications of probability? Let us begin our attempt to answer this question by dealing first with a theory I hold to be completely unsatisfactory.

The subjectivistic theory of probability (SUB) just will not do.

An adequate view of probability theory must, I think, be a realistic one; not in the sense that probabilities must be eternal Platonic objects, but in the sense that Putnam describes and attributes to Dummett:[4]

> I am indebted to Michael Dummett for the following very simple and elegant formulation of realism: A realist (with respect to a given theory or discourse) holds that (1) the sentences of that theory or discourse are true or false; and (2) that what makes them true or false is something *external* – that is to say, it is not (in general) our sense data, actual or potential, or the structure of our minds, or our language, etc. Notice that, on this formulation, it is possible to be a realist with respect to mathematical discourse without committing oneself to the existence of 'mathematical objects'. The question of realism, as Kreisel long ago put it, is the question of the objectivity of mathematics and not the question of the existence of mathematical objects.

In the case of probability theory, if Smith says the probability of throwing a Five is 1/6, and Jones say it is 1/2, their dispute is just as genuine and the correct answer is just as objective as if they were directly arguing about whether or not the die is physically loaded. The one who is right – the one who *knows* the *real* probability – not only possesses the epistemological virtue of knowing the truth, he also has the practical advantage of being in a position to improve his economic situation at the expense of the other fellow. It is the fatal flaw of the subjectivistic theory that it pretends that one person's probability is just as good as the other's and that no objective consideration can choose between them, when it is perfectly plain that one will lose his shirt *because he is wrong about the probability.*

It is of course true that each person has various *opinions* about probability, and it is useful for the SUB theorists to identify the minimal conditions that each must obey in order for those views to be consistent. But as Putnam has also pointed out,[5] we ask more of mathematical systems than *mere* consistency; we ask also that they truly describe the world.

Thus far we have not settled on a view of probability. It may be metaphysical or epistemological, physical or conceptual. We have only established that it must appeal to some standard of truth or objectivity outside the opinions of humans.

Now let us dispose of the Classical theory.

To oversimplify brutally: what the CTP amounts to is the discovery and elaboration of the most important rule of thumb in probability theory, the Principle of Indifference.

As I indicated earlier, I think the fundamental successes of the Principle of Indifference are due to the fact that its application tends to have the features of a Random Guess. This accounts for the puzzling fact that the Principle works best in a situation of ignorance; it is not that the universe conforms itself to our expectations, it is just that our probability of success in a random guess is a fixed and equal-valued function of the alternatives we judge to be equipossible, *whether or not those alternatives are equiprobable in some objective sense.* It is not the occurrence of the *events* which need be equiprobable for the Principle to hold, but the occurrence of *successful guessing.* (This situation has been obscured by the fact that many of the most important uses of probability theory involve alternatives which are, or are intended to be, equiprobable, so that the Principle of Indifference tends to be externalized and reified – this is an understandable but regrettable error.)

Again let me say that the Classical theorists were intellectual giants, and I would never wish to denigrate their accomplishments. I am only arguing that most of their writing was non-philosophical, and their one great insight is not a philosophical explanation of what probability is, but a practical rule for evaluating and acting on probabilities. It is a tremendous rule, but it won't do as a theory. (It will do very nicely as a part of a theory, of course, and is in fact compatible with either of the remaining types, since the justification of a random guess can be either a logical a priori insight or calculation, or the observed fact that random guesses exhibit a stable frequency of success.)

Now we come down to the great theoretical divide. We are left with only two major players on the field – the a priori and the relative frequency theories. At this point there is so much to be said for and so many supporters of each view that it would be perfectly reasonable just to accept them both. Carnap, of course, does just that, with his view that probability$_1$ and probability$_2$ are two different concepts, two different equally valid explicanda for probability theory, and that in normal usage the term 'probability' is ambiguous between them unless made clear by context. In a similar fashion, Hacking speaks of epistemic and aleatory probabilities.[6] And others have held that it is not possible to combine in one theory the two equally valid views of probability as an 'outer' property of things and events and as an 'inner' property of propositions and knowledge. Let me say a few words about this problem.

First off, I think that it is philosophically useful (or, better, analytically useful) to distinguish senses of terms and shadings of meanings. It helps us to get clearer on what we're talking about and to understand the variations in what others are talking about. But it is a nobler philosophical goal and a far sounder scientific goal to unify our meanings, fix the significance of our terms, and bring order out of chaos. I therefore think that we cannot tolerate ambiguity – even systematic ambiguity – in so crucial a term as 'probability.' Instead, I think it would be more helpful if we view 'probability' as what Putnam calls a 'law-cluster term'.[7] A law-cluster term is one which enters into several different laws, deriving part of its significance from each. Consequently, there can be no unchangeable analytic definitions of such terms, as shifts in the direction of science will magnify the importance of some 'definitions,' invalidate others. (This is how 'Atoms are indivisible particles' and 'Whales are the biggest fish in the sea' began as analytic definitions of simple terms, evolved into laws involving law-cluster terms, and finally became falsehoods as the other laws became both more important than and incompatible with these statements.)

I will go further – not only is 'probability' a law-cluster term like 'atom,' it is, like 'gravity,' a term in a state of flux and tension, in search of a paradigm. In the case of 'gravity,' it may be that we will plump for the view that gravity is (something like) the curvature of the space-time metric which occurs in the vicinity of masses and which therefore causes such masses to tend to move closer (be closer?) together. If so, all sentences beginning 'Gravity is a (the) force...' which were once true – perhaps even analytically true(?) – will then be false.

Now in the case of probability, I think that the *purpose* of probability theory is to bring the degree of uncertainty in our world view as much as possible into congruence with the uncertainty in the empirical world. I use the word 'uncertainty' here because it conveys much of the traditional notion of probability, but also because it is now in just the state of physical vs. epistemological indecision that probability theory languishes in (witness the Copenhagen Interpretation of the Heisenberg Uncertainty Relations[8]).

For those who insist on a definition, I suppose I would start with 'Probability is the measure of uncertainty in a given part of our description of the world.' But this is not a complete definition. It is rather the kind of thing that John Rawls would call a *concept* of probability – a rather vague, ambiguous, or, especially, underdetermined notion which then achieves specificity by the introduction of a *conception* of probability.[9] The concept is normally compatible pairwise with each of two or more conceptions which are mutually *in*compatible. Thus, to accept the definition of a concept is just to set the terms of our true problem, which is to specify the best conception.

And this is just the situation I find probability theory to be in. If we accept my definition of the concept of probability – the measure of uncertainty in a given part of our description of the world – it is immediately obvious that such uncertainty could be the result of the incompleteness and inadequacy of our knowledge, or it could be the result of randomness or indeterminism in the world itself. Many of the laws invoking the term 'probability' are concerned with one type of uncertainty, many are concerned with the other, and some are not clearly one or the other.

It is important, but not sufficient, to distinguish these conceptions of probability. Eliminating confusions and equivocations is a necessary prolegomenon to finding the truth – but it does not *constitute* finding the truth. That is why I reject any principle of tolerance which says that different theories should be 'equally' accepted. The aim is not to distinguish different conceptions but to find the best (true?) conception. Since the aim of probability is to bring the uncertainty in our descriptions as much as possible into congruence with the uncertainty in the world, it seems to me that the most important question in probability theory is the physical/metaphysical/cosmological question 'Is there genuine chance or randomness in the world?'

This question lies at the conceptual heart of probability theory because if it is answered in the negative we need deal with only one type of uncertainty (roughly, epistemological uncertainty), while if it is answered in the affirmative, we must also deal with physical chance. If the latter is true, we might be forced back to Carnap's two explicanda (in which case we need even stronger verbal differentiation than his probability$_1$/probability$_2$), or we might yet hope for a unified theory dealing with some common element in each. In either case, it would be conceptually quite different from just trying to develop a system of partial belief or judgment in a world of deterministic regularity and fixed truth values.[10]

Strangely enough, the grand division between AP and RF theories does not turn precisely on this important point. Although von Mises originally insisted on a form of 'randomness', subsequent RF theorists have developed rules and techniques which lead to success or failure quite independently of whether or not true randomness exists. At the same time, their official position implies that there can be no such thing as the probability that an individual radioactive atom will decay in the next ten minutes, even if such an event turns out to be the purest possible case of genuine physical chance.

And so I must finally reject the RF interpretation of probability. Basically I accept the charge that RF theorists *in practice* confuse the evidence for a probability and the manifestation of a probability (both of which are frequently frequencies) with the fact of the probability itself. Whatever probability might be, it is surely not the limit of a relative frequency in an infinite series of empirical events. In the first place, there may not be any such series *at all* (if the universe is finite, as seems quite possible). But any theory which implies that there may not be probabilities is surely false, because it is quite plain that there *are* probabilities (again, not in the Platonic sense of eternal universal objects but in the Putnamian sense that probability theory *works*). In the second place, an adequate theory of probability must deal with the large number of contexts where probability applies to individual events or evidentiary relationships between propositions (and not in the strained and fictitious sense that Reichenbach tries to invoke).

In the end, I think that probability theory is a part of logico-mathematics which lies far closer to the 'a priori' end of the spectrum, nestling up against deductive logic and geometry, than to the 'relative frequency' end, in bed with insurance companies and quantum mechanics. But Keynes is too intuitive and Carnap too abstract to capture exactly what I mean.

I share the Harvard view of theory and the a priori, conceived in C. I. Lewis's conceptual theory of the a priori, tempered by the fire of Quine's, White's, and Kuhn's searching challenges, and brought to maturity in Putnam's philosophy of mathematics and Rawls's ethics. I hold no statements or principles to be eternal and unrevisable, and no 'facts' to be brute and incorrigible. Yet it *does* matter what we believe, and truth is both preferable to error and attainable by human beings, because there *is* a real world out there to which we must conform our theories or suffer the consequences.

In the case of probability theory, the anti-foundationalist mathematicians are correct that Kolmogorov's axiomatization and the theorems derivable therefrom are at least as secure as number theory – they need no peculiarly mathematical foundation. And those who worry only about the practicalities of the matter can cheerfully say of probability theory what Putnam said of the differential and integral calculus: 'The point is that the real justification of the calculus is its *success* – its success in mathematics, and its success in physical science'[11] – probability theory needs no further pragmatic foundation. But the point Putnam intended to make was not just the anti-Platonic one that mathematics could be quasi-empirical and need not depend on eternal objects; he also wanted to stress the genuinely realistic position that some mathematical theories – like some physical theories – are better than others. Some are true and some are false (of each type) and we must look to the world to see which is which, no matter how consistent and elegant they may be.

So now we see the other connection I wish to establish: probability theory is closer to logic than von Mises would have it, but *logic is closer to science than Keynes ever dreamed.*

My paradigm case of good probabilism is that of the Bose–Einstein statistics. Bose and Einstein *made up* an a priori model, using, in fact, the Classical rule of thumb to distribute particles among state-descriptions. But then they checked their mathematical theory against reality, using, in fact, the relative frequency of distributions as evidence of what the probability really was. And they discovered that their model applied to some particles but not to all. Finally, Born's quantum-theoretical analysis gave a physicalistic difference between those which did and those which did not obey Bose–Einstein statistics.

Probability theory is intended as a mathematical description of the world. Its goal is to bring the uncertainty in our world view as closely as possible into congruence with the uncertainty in the world. Our mathematicians have done a great job of constructing and elaborating mathematical systems and theorems towards that end. Our scientists have made considerable progress in discovering which physical systems obey which models. But no adequate philosophical explication of probability theory exists as yet, nor can one do so until we learn a good deal more logic and a good deal more physics than we presently know. We may never be sure we have it right, perhaps, until we possess a general theory of rationality and know for certain whether or not God plays dice with the Universe.

NOTES

I WHAT IS PROBABILITY?

1 C. I. Lewis, *An Analysis of Knowledge and Valuation*, p. 320. Hereafter cited as *AKV*.
2 J. M. Keynes, *A Treatise on Probability*, p. 8.
3 This is not an uncontroversial suggestion. Hans Reichenbach believes that 'the probability concept...can be studied successfully only within the realm of its scientific application.' (H. Reichenbach, *The Theory of Probability*, p. 5. Hereafter cited as *TOP*.) Ordinary language philosophers would presumably take the contrary view.
4 This is a paraphrase of what Nelson Goodman said about both deductive and inductive inference in *Fact, Fiction, and Forecast*, p. 64.
5 Ibid., p. 66.
6 *TOP*, ch. 3, pp. 45–123.
7 This requirement is rejected by some theorists – especially mathematicians – on the grounds that probability is neither more nor less than a unit-sum additive measure-function which is pragmatically applied to events. In this view, no further definition or interpretation of 'probability' is either possible or necessary. Some authorities list this as an additional theory, definition, or interpretation of probability, but I ignore it on the grounds that an empty theory is no theory at all.
8 Indeed, the theoretical similarity is so great that I included the CTP as merely a historical portion of the AP group of theories in my doctoral dissertation, 'Probability and certainty in C. I. Lewis's epistemology'.
9 R. Carnap, *Logical Foundations of Probability*. 1950 edn, p. 19.
10 *AKV*, pp. 266, 269–70.
11 'Probability and certainty', pp. 102–208.
12 E. Nagel, *Principles of the Theory of Probability*.
13 Ibid., pp. 18–19. Paragraphing and numerals added for clarity.
14 My (and Carnap's) arguments against the subjectivist and psychologistic construal of Classical theories will be found in the next chapter.

15 H. E. Kyburg Jr and H. E. Smokler (eds), *Studies in Subjective Probability*, p. 4.
16 H. E. Kyburg Jr, *Probability and Inductive Logic*.
17 I. J. Good, 'Subjective probability as the measure of a non-measurable set,' p. 319.
18 Ibid., pp. 319–20.
19 G. H. von Wright, *The Logical Problem of Induction*, pp. 150–3.
20 M. Black, 'Probability,' *Encyclopedia of Philosophy*, vol. 6, pp. 474–7, summarized and paraphrased.
21 T. L. Fine, *Theories of Probability*.
22 Ibid., pp. 8–10.

II THE CLASSICAL THEORY OF PROBABILITY

1 I. Todhunter, *A History of the Mathematical Theory of Probability From the Time of Pascal to That of Laplace*.
2 J. M. Keynes, *A Treatise on Probability*, 1921 edn, p. 432. Hereafter cited as *Treatise*.
3 F. N. David, *Games, Gods and Gambling*, p.x. Hereafter cited as *GGG*.
4 Actually, it is now possible to supplement Todhunter with David's much more readable early history, which is just as scholarly and much less technical. One can only hope that she will continue her excellent work in the near future by going forward beyond De Moivre (1667–1754).
5 *GGG*, p. 4. This entire paragraph is based primarily on the first three chapters of that work.
6 Suetonius, *Lives of the Caesars*, quoted in *GGG*, p. 8.
7 E. Nagel, *Principles of the Theory of Probability*, p. 7. Hereafter cited as *Principles*.
8 L. E. Maistrov, *Probability Theory: A Historical Sketch*, p. 5.
9 *GGG*, ch. 4, 5.
10 See p. 2 of Todhunter (op. cit.) for his opinions and those of Montmort and Libri.
11 Maistrov, op. cit., p. 18.
12 *GGG*, p. 58.
13 Ibid., ch. 7.
14 Maistrov, op. cit., p. 30. Maistrov nevertheless gives Galileo more importance than do other historians.
15 P. S. de Laplace, *A Philosophical Essay on Probability*, pp. 167, 185.
16 Todhunter (op. cit., p. 7) sympathetically quotes three such descriptions. For a typical modern example see L. Hogben, *Mathematics in the Making*, p. 250.
17 W. Kneale (*Probability and Induction*, p. 123), Hogben (op. cit., p. 250), and several sources cited by Maistrov (op. cit., pp. 40–3), think that 'the Chevalier de Méré's problem' concerned the probability of throwing Double Six in 24 casts of two dice. Todhunter (op. cit., ch. II) and David (*GGG*, ch. 9) include this problem but tend to stress instead the importance of 'The Problem of Points' (division of stakes for an

incomplete game) which was *also* apparently transmitted to Pascal by the Chevalier de Méré.

18 See, for example, F. N. David, 'Dicing and Gaming,' p. 15, and M. G. Kendall, 'The Beginnings of a Probability Calculus,' p. 19, both in E. S. Pearson and M. G. Kendall (eds), *Studies in the History of Statistics and Probability*.

19 Maistrov, op. cit., p. 55; *GGG*, p. 110.

20 *Treatise*, pp. 41, 81.

21 J. M. Keynes, a man of not inconsiderable scholarship, seems inexplicably unaware of Huygens's priority in this regard, as a footnote on p. 311 of his *Treatise* attributes the first expression of this idea to a later work of Leibniz (1678).

22 *GGG*, p. 115. This view also appears in Maistrov, op. cit., p. 49 (apparently paraphrased from David's account), and is in agreement with Todhunter, op. cit., p. 25.

23 So called because D. Bernoulli's original discussion appeared in the *Papers of St Petersburg Imperial Academy of Sciences*, 1738.

24 The chief puzzle is that the mathematical expectation of the Petersburg game seems to be infinite, since the rules allow for the game to continue indefinitely until Heads shows up – yet no rational person would pay even a moderately large stake to play the game.

25 *Treatise*, pp. 41, 81. R. Carnap agrees. (*Logical Foundations of Probability*, 1950 edn, p. 48.) Hereafter cited as *LFP*.

26 Besides Todhunter, various sources including Maistrov and David (who unfortunately stops her narrative just before Daniel Bernoulli) have been used for this brief account of the Bernoullis.

27 The form, though not the explanation, of this equation is taken from W. Feller, *An Introduction to Probability Theory and Its Applications*, p. 124. As Maistrov notes, this formula 'actually appears nowhere in Bayes' writing. It is Laplace who was responsible for this name' (op. cit., p. 100).

28 Hogben, op. cit., p. 269.

29 *Treatise*, p. 365n.

30 Maistrov, op. cit., p. 123. He then goes on to quote a Russian text which takes this view of d'Alembert.

31 Todhunter, op. cit., p. 464.

32 *Collier's Encyclopedia*, 1966, vol. 14, p. 320.

33 Laplace, op. cit., pp. 6–7.

34 James (Jacob) Bernoulli, *Ars conjectandi*, excerpted in J. R. Newman (ed.), *The World of Mathematics*, $i = 3$, pp. 1452–5, brackets in Newman, parentheses in original.

35 *LFP*, p. 50.

36 See Kneale, op. cit., p. 124, and *Principles*, pp. 18, 44.

37 *LFP*, pp. 47–50.

38 *GGG*, p. 26

39 Maistrov, op. cit., pp. 5–8.

40 David suggests, for example, that Cardano noticed 'that if all the elements of this set are of equal weight then the ratio of the number of

favorable cases to the total number of cases gives a result in *accordance with experience.*' (*GGG*, p. 58, emphasis added.)

41 *Treatise*, p. 41.
42 Laplace, op. cit., p. 56, where the subject is actually a biased coin, but the theoretical point is clear.
43 Ibid., p. 57.
44 J. Bernoulli, op. cit., p. 1453.
45 Laplace, op. cit., pp. 141–2.
46 Todhunter, op. cit., p. 594.
47 See pp. 30, above.
48 This is properly called 'Bernoulli's Limit Theorem' to distinguish it from the more basic 'Bernoulli's Theorem' which specifies the probability that *F* will fall in the desired range.
49 See section 'Source of initial probability in actuarial cases,' pp. 31–4.
50 'A table of mortality is then a table of the probability of human life. The ratio of the individuals inscribed at the side of each year to the number of births is the probability that a new birth will attain this year.' Laplace, op. cit., p. 142.
51 Ibid., p. 19.
52 Feller, op. cit., pp. 123–4.
53 Laplace, op. cit., p. 19. The reader is reminded that a bet of 1,826,214 to one corresponds to a probability of success of 1,826,214/1,826,215, which is

$$\frac{1,826,213 + 1}{1,826,213 + 2} \quad \text{or} \quad \frac{N + 1}{N + 2}.$$

In trying to roll a Five, for example, with a fair die, the probability of success is 1/6, but the *odds* are 1 to 5 for you since *one* possible success (win) corresponds to *five* possible failures (losses).
54 This is the interpretation which Carnap accepts (*LFP*, pp. 47–51) and which I argued for in my dissertation, 'Probability and certainty in C. I. Lewis's epistemology.'
55 See, for example, the passage which begins 'If all events...' on p. 1455 of Newman, op. cit.
56 Idem. See also Maistrov, op. cit., p. 67.
57 P. R. Montmort, *Essai d' Analyse sur les Jeux de Hasard*, Paris, 1705, quoted in *GGG*, p. 150. And see Maistrov, op. cit., p.77.
58 Laplace, op. cit., p. 4.
59 For example, quantum mechanics holds that the present state of a particle determines its future state only according to a probability distribution – not rigidly and predictably. A given particle in my body might 'decide,' then, to head straight up rather than go along with the rest. In principle, even, they might all decide to go straight up, so that I would levitate, in apparent violation of the Law of Gravity. The odds against such an occurrence are so tremendous that we can be virtually certain it would never happen to anyone even in many repetitions of the entire life of the universe. The puzzling thing is that quantum mechanics seems to allow at least a metaphysical possibility that such events might

occur by chance. That is why many scientists (including, most notably, Albert Einstein) have been unsatisfied with quantum mechanics and have hoped for the development of a new physics which would not rely on the inherently probabilistic psi-function as the basic description of reality.

60 Laplace, op. cit., p. 6.

61 If this principle is regarded as generally true, and if one accepts a utilitarian view of economics or morality, there results a very powerful argument in favor of mechanisms for the redistribution of wealth from the rich to the poor. Since the rich man will lose less than the poor man will gain in the transfer of a given amount, it follows that the happiness of the poor man will outweigh the sadness of the rich, and thus there will be more total happiness than before. Anyone committed to maximizing total happiness, then, should always take from the rich and give to the poor.

62 Laplace, op. cit., p. 24.

63 Maistrov, op. cit., pp. 41–2.

64 *GGG*, ch. 6.

65 See Todhunter, op. cit., p. 100.

66 Laplace, op. cit., p. 196.

67 Kyburg calls it 'the most notorious principle in the whole history of probability theory' (*Probability and Inductive Logic*, p. 31), while C. I. Lewis refers to it as 'that *bête noir* of clear thinking' ('Review of John Maynard Keynes's *A Treatise on Probability*,' p. 182).

68 H. Reichenbach, *The Theory of Probability*, p. 353.

69 Kyburg especially sees Laplace as attempting this. (Kyburg, op. cit., p. 31.)

70 Laplace, op. cit., p. 6.

71 Keynes, rather idiosyncratically, requires that the Principle of Indifference be applied only to 'ultimate' cases, by which he means to exclude not only 'greater than Two' but also Odd and Even, since the latter are also reducible to sub-cases (though in this case to an equal number of sub-cases). Most analysts do not follow him in this requirement, but it has the virtue of excluding all the forbidden cases and making us aware of which are which by analyzing each case as far as possible. Of course legitimate divisions such as Odd and Even can be reconstructed from the ultimate equipossibility of 1, 3, 5, 2, 4, 6. (See *Treatise*, ch. IV.)

72 C. I. Lewis, *An Analysis of Knowledge and Valuation*, p. 307. Hereafter cited as *AKV*.

73 *Treatise*, ch. IV.

74 *Principles*, p. 47.

75 This is the conclusion reached by Lewis on p. 314 of *AKV*.

76 Named after Joseph Bertrand (1822–1900), who first propounded it. Not to be confused with 'Bertrand's Paradox' which deals with geometrical probability and will be discussed later.

77 This paradox is described in Kyburg, op. cit., pp. 34–5.

78 A somewhat different version of Bertrand's Paradox is in ibid., pp. 36–7.

79 J. von Kries, *Die Prinzipien der Wahrscheinlichkeitsrechnung, Eine logische Untersuchung*, Freiburg, 1886, p. 24, cited on p. 45 of *Treatise*.

80 T. L. Fine, *Theories of Probability*, p. 170.

81 Ibid., p. 167.

82 *Principles*, p. 47, emphasis in original.

83 Kyburg, op. cit., p. 33. The point is clear and well taken, but there seems to be an error in the example. There are only 168 primes in the first 1,000 natural numbers, with each division of 100 containing, in order, 25, 21, 16, 16, 17, 14, 16, 14, 15, 14. This relative frequency of 0.168 is certainly a far cry from $6/\pi^2$, which is about 0.609.

84 R. von Mises, *Probability, Statistics, and Truth*, p. 70. Hereafter cited as *PST*.

85 A. J. Ayer argues this point in *Probability and Evidence*, p. 28. See also Keynes, *Treatise*, p. 363 for an example of dice rejected as 'ill-made'. (The reader may wish to wait until the chapter on a priori theories, when this example will be discussed.)

86 Bernoulli, op. cit., pp. 1452–3.

87 Poisson and Quetelet were especially prone to this lapse, and Laplace was not immune.

88 Kneale makes this point on p. 140 of *Probability and Induction*.

89 Von Mises considers these to be the ordinary base-10 numbers, mostly very large, which fortuitously can be written by a combination of up to 100 '0's and '1's. He knows there are 2^{100} of them because they are in one-to-one correspondence with a permutation which he knows has 2^{100} elements. A modern reader will perhaps be more comfortable considering them to be the series of numbers which result if, starting with 0, we *count* each possible case in *binary arithmetic*. This is easily seen to be the series from 0 to 2^{100-1} or the first 2^{100} binary numbers.

90 *PST*, p. 108.

91 Ibid., p. 115.

92 J. Newman, 'Commentary on the Law of Large Numbers,' in Newman, op. cit., p. 1449.

93 Ibid., p. 1450.

94 *Principles*, passim.

95 *LFP*, pp. 47–50.

96 Ayer, op. cit., p. 28.

97 See next chapter on a priori theories of probability.

98 Feller, op. cit., pp. 40–1, emphasis in original.

99 The phase space is multi-dimensional, so that each cell has coordinates assigning to it a momentum as well as a position in physical space. Assignment of a particle to a cell thus determines both the particle's position and its momentum.

100 Feller, op. cit., pp. 40–1.

101 M. Born, 'Einstein's statistical theories,' pp. 174–5.

102 The meanings of these terms and the theoretical significance of this choice will appear in the discussion of Carnap's theory in the next chapter.

103 Kyburg, op. cit., p. 38.

104 R. W. Marks, *The New Mathematics Dictionary and Handbook*, p. 38.
105 E. B. Uvarov, D. R. Chapman, and A. Isaacs, *A Dictionary of Science*, pp. 239–40.

III A PRIORI THEORIES OF PROBABILITY

1 It is, of course, a feature shared by the Classical school. For this reason, I included Classical theory in the class of AP theories in my doctoral dissertation, 'Probability and certainty in C. I. Lewis's epistemology.' I still think that classification is more theoretically correct, but the historical importance, unity, and recognition of the Classical school convinced me to treat them separately in this more comprehensive work.

2 Kyburg's name 'Degree of Entailment Theories' (in H. Kyburg, *Probability and Inductive Logic*) is presumably intended as neutral between Carnap's 'degree of confirmation' and Keynes's 'degree of rational belief.' It is certainly a descriptive term, but philosophers may reject 'partial entailment' as one more contradiction they can do without.

3 From the (unattributed) frontispiece of J. M. Keynes, *Essays in Biography*.

4 J. M. Keynes, *A Treatise on Probability*, 1962 edn. Hereafter cited as *Treatise*.

5 R. Carnap, *Logical Foundations of Probability*, 1950, p. 31, hereafter cited as *LFP*. But Keynes himself attributes this insight to L. M. Kahle in his *Elementa logicae Probabilium methodo mathematica in usum Scientiarum et Vitae adornata*, Halle, 1735, and notes 'casual statements' to that effect in Boole, Bradley, and Laplace. *Treatise*, p. 901.

6 Norman M. Martin, article on Rudolf Carnap in the *Encyclopedia of Philosophy*, vol. 2, p. 25. Similarly, Carl G. Hempel calls him the 'leading representative of logical empiricism' in his *Collier's Encyclopedia* article on Carnap (p. 457).

7 Martin, op. cit., p. 31.

8 *LFP*, pp. 52–161.

9 *Treatise*, pp. 3–4.

10 Idem.

11 Ibid., p. 40.

12 This argument appears in bk I, ch. III, of the *Treatise*.

13 Readers with a logico-mathematical bent are no doubt frustrated by this talk about the value of a relation. Keynes's probability-relation is of course not a relation in the logical sense of being a dyadic propositional function which is either true or false of each ordered pair of propositions. Yet neither is it a function which takes propositions as arguments and unambiguously assumes a numerical value. It is as though we were talking about 'the deductive relation between P & Q' and offered '$D(P,Q)$' as an abbreviation for the words in quotes. Then we could say 'When "$D(P,Q)$" is entailment we will agree that "$D(P,Q) = 1$" is true and refers to that state of affairs; similarly with contradiction and "$D(P,Q) =$

O".' Then we could indeed define a function $DF(P, Q)$ which has propositions as arguments and 0 and 1 as possible values, depending on whether $D(P, Q)$ is contradiction or entailment. But there would remain areas in which '$DF(P, Q)$' is undefined and has no value, while '$D(P, Q)$' *is* defined (as 'the deductive relation between P and Q') and may have significant meaning (e.g., reverse implication, which is not relevant to $DF(P, Q)$). If we now change the function's name from '$DF(P, Q)$' to '$D(P, Q)$,' we will have a defined symbol which is *ambiguous* between naming a function and meaning the deductive relation between P and Q. This, I think, is the logical status of 'a/h,' when 'deductive relation' is replaced by 'probability relation.'

14 *Treatise*, p. 8.
15 Ibid., pp. 3–4.
16 Ibid., p. 38.
17 Ibid., pp. 12–14.
18 In the section on 'The epistemological status of P.'
19 *Treatise*, p. 65.
20 C. I. Lewis, 'Review of John Maynard Keynes's *A Treatise on Probability*,' p. 182.
21 E. Nagel, *Principles of the Theory of Probability*, p. 62.
22 *LFP*, p. v.
23 Ibid., p. xiv. Parentheses in original.
24 Ibid., p. 164.
25 Ibid., p. ix.
26 Ibid., p. 169: 'e' represents the evidence.
27 Carnap recognizes that equating probability$_1$ to such an estimate of a frequency is a 'circular procedure', but he says it is not a vicious circle since he is not here establishing a 'system of definitions', but only seeking 'a clarification of certain concepts as explicanda.' He promises to remove the circularity when he constructs a formal system in a later section (pp. 172–3). But all that he does in the latter place is to define estimates in terms of degrees of confirmations. The dependence is then one-way rather than circular, but the 'estimate of a frequency' is then explicitly dependent and secondary rather than fundamental. (*LFP*, p. 525.)
28 *LFP*, p. 294.
29 Ibid., p. 71.
30 For expositional clarity and typographical simplicity I have taken the liberty of changing some of Carnap's German and Greek symbols into English, even in quotations. These changes are straightforward enough not to confuse anyone already familiar with the system, and, of course, have no systematic import at all.
31 *LFP*, p. 79.
32 Note that in this sentence we adopt the mathematical convention of sometimes allowing such object-language symbols as 'Ma & Na' to serve as *names of themselves* in the meta-language. This practice is common to most mathematicians and many logicians (including Carnap) and is particularly helpful when these symbols are to be combined with such meta-linguistic symbols as 'R.' I trust that a judicious employment of

this procedure will not confuse a normal reader and will serve as an innocent source of pleasure for those who get their jollies out of counting use-mention 'mistakes.'

33 In Carnap's essay. 'On inductive logic,' a footnote on p. 45 cites passages *4.4, *4.26, *5.101, and *5.15 in Ludwig Wittgenstein's *Tractatus Logico-Philosophicus.*

34 G. H. von Wright, in his book, *The Logical Problem of Induction*, includes Wittgenstein, Halperin, Czuber, Waismann, and Drobisch in this school, but Friedrich Waismann's work is perhaps most associated with logical range theories other than Carnap's.

35 *LFP*, p. 295.

36 Ibid., pp. 564–5.

37 $c^\dagger$ does learn from experience in those cases where the acquired evidence, e, says something about the object in question. In our example, if e includes Mb, $c^\dagger(h, e)$ increases to $1/2$.

38 *LFP*, p. 116 (*D27–1*).

39 This is true in *LFP*. In his later work Carnap no longer argues that one c-function is satisfactory in all cases but tries to develop a theoretical description of an infinite continuum of c-functions. We are here only concerned with the system of *LFP* – those interested in the later work should consult R. Carnap, *The Continuum of Inductive Methods.*

40 Strictly speaking, the system of *LFP* is not applicable to this problem at all, since the predicates are not independent in the required sense (a roll can't be both a Five and a Four). But in his later work Carnap revised the system to deal with 'families' of predicates where one and only one member of the family can be predicated of an individual ('red' or 'green' but not both, for example). John Kemeny discusses the application of this method to the throw of a die in Section IV of 'Carnap on probability and induction.'

41 *LFP*, p. 501.

42 *Treatise*, p. 108.

43 Ibid., p. 98.

44 Ibid., chs XVIII–XXII.

45 The 'Principle of Limited Independent Variety' is a modern name for what Keynes himself called the 'Inductive Hypothesis' which basically asserts that the relevant system is finite, from which it follows that any two properties have some finite a priori probability of being associated with each other.

46 'Let the reader be clear about this. To argue from the mere fact that a given event has occurred invariably in a thousand instances under observation, without any analysis of the circumstances accompanying the individual instances, that it is likely to occur invariably in future instances, is a feeble inductive argument, because it takes no account of the Analogy.' *Treatise*, p. 407.

47 This example is discussed on p. 397 of the *Treatise.*

48 Keynes approves of and partially adopts the methods of W. Lexis for evaluating significant dispersion. *Treatise*, ch. XXXII.

49 To meet *LFP*'s requirement of independence, it will be necessary to

assume that the individuals live in different cities, so that the election of one as mayor won't preclude the simultaneous election of the others. (Alternatively, '*M*' could be 'elected Mayor at some time.')

50 *T35-2c, LFP*, p. 139.

51 To see this quickly, change the lexical ordering of the *Z*s so that '*Ms*' is the first atomic sentence. Then '*Ms*' is true in the first 256 *Z*s, '*t*' is true in all 512 *Z*s, and, since the *Z*s are symmetrically weighted, $m^*(Ms, t)$ is 1/2.

52 *LFP*, p. 125.

53 This gives us an additional method of specifying structure-descriptions. $STR_1 = 3Q1, STR_2 = 2Q1 \& 1Q2$, etc. This is the method of Q-numbers. (*LFP*, pp. 134–7.)

54 *LFP*, p. 127.

55 This method is described on p. 569 of *LFP*.

56 See next section on 'Probability of repetitive kinds of events.'

57 Given extreme simplifying assumptions and modifying c^* for one family of predicates.

58 See below, pp. 132–3.

59 *LFP*, p. 171.

60 *Treatise*, p. 413.

61 Ibid., p. 414.

62 Ibid., p. 362.

63 Ibid., p. 363.

64 A. J. Ayer, *Probability and Evidence*, p. 28.

65 I have expounded on Keynes's reply to Wolf because it seems to me to illustrate a vital point concerning a priori theories in general. But in fact there are at least two other reasons why subsequent evidence cannot refute the a priori assignment of probabilities. The reader might try to identify these before we discuss them in the section on common features of AP theories.

66 *Treatise*, pp. 286–7.

67 Ibid., pp. 283–7.

68 Ibid., p. 286.

69 Ibid., p. 289.

70 The definition frequently found in scientific and mathematical texts, whereby a phenomenon is said to be random if and only if it is pragmatically impossible for us to predict it is practically and even epistemologically adequate, but of course it has no metaphysical significance.

71 See 'Intellectual autobiography,' in P. A. Schilpp (ed.), *The Philosophy of Rudolf Carnap*, pp. 75–6, for Carnap's hints concerning his unpublished work in this area. The above definitions are illustrative only and not intended precisely to represent Carnap's views.

72 Ayer, op. cit., p. 39.

73 Schilpp, op. cit., p. 15.

74 *Treatise*, pp. 6–7.

75 Ibid., pp. 130–1.

76 Ibid., ch. XXXI.

77 *LFP*, p. 332.

78 Carnap endorses this procedure on p. 84 of *The Continuum of Inductive Methods*.

79 A. J. Ayer has claimed that it is and has pressed the argument especially strongly in *Probability and Evidence* and 'The conception of probability as a logical relation,' 1957, pp. 12–17. The latter essay is reprinted in M. H. Foster and M. L. Martin (eds), *Probability, Confirmation, and Simplicity*, where, on pp. 20–1, the editors discuss Ayer's criticism and recommend essentially the reply I present here.

80 Both Keynes and Carnap define relevance in this way (although neither uses the statistician's term 'stochastic independence'). See *Treatise*, pp. 55, 146f., and *LFP*, p. 348.

81 N. Arley and K. R. Buch, *Introduction to the Theory of Probability and Statistics*, p. 19.

82 *LFP*, p. 211, *sic*.

83 Idem.

84 Ibid., pp. 208–19

85 Ibid., p. 20.

86 Foster and Martin, op. cit., p. 37.

87 *LFP*, p. 30.

88 *Treatise*, p. 32.

89 Ibid., pp. 32–3.

90 Ibid., p. 245.

91 I say 'generally accepted' because such heterodox theorists as Condorcet and Borel argue that very probable events do always occur and that very small probabilities are really equal to zero.

92 *Treatise*, p. 131.

93 Ibid, p. 245.

94 *LFP*, pp. 149–50.

95 Schilpp, op. cit., p. 978.

96 *LFP*, p. 2. In 'My basic conceptions of probability and induction' he further delimits his aim as 'an explication of the concept of logical probability...'. Schilpp, op. cit., p. 967.

97 *LFP*, p. 5.

98 It should be noted that there are reasonable grounds for alternative interpretations of Carnap's views. By concentrating on *The Continuum of Inductive Methods* and certain passages of *LFP*, it is possible to view Carnap as espousing the pragmatic justification of inductive methods. This approach is taken by J. W. Lenz in his article 'Carnap on defining "degree of confirmation",' which was written before the publication of the Schilpp volume. In it, Lenz emphasizes 'performance' as the criterion of choice for inductive methods, and notes that we cannot know that a c-function will perform well in the future, we can only predict that it will do so, and such prediction requires c or some other inductive method to be assumed. We *can* know that c has performed well in the past, but this is no guarantee for the future unless we again assume an inductive rule. Notice that these difficulties are avoided if our requirement is that c agrees with our intuitions, since we always appeal only to present intuitions.

99 *Treatise*, p. 3.

100 Ibid., p. 12.

101 Ibid., p. 131.

102 Ibid., p. 133.

103 This phrase almost appears in Keynes, when, on p. 261 of *Treatise*, he writes in lower case about 'the hypothesis of the limitation of independent variety' when he is clearly referring to the Inductive Hypothesis.

104 *Treatise*, p. 258.

105 Ibid., p. 263.

106 Idem.

107 Ibid., p. 262.

108 Ibid., p. 264.

109 Putnam's suggestion was in a private conversation. It is strongly supported by Keynes's acknowledgment of Moore's influence in the Preface to *Treatise* (p.v) and especially by his description of the overwhelming effect of Moore's system on his beliefs as an undergraduate, which is to be found in the charming 'My early beliefs,' in *Two Memoirs*, p. 93 where he says of Moore: 'The large part played by considerations of probability in his theory of right conduct was, indeed, an important contributory cause to my spending all the leisure of many years on the study of that subject: I was writing under the joint influence of Moore's *Principia Ethica* and Russell's *Principia Mathematica*.'

110 Another argument in favor of Keynes's position develops out of his discussion of 'unknown probabilities.' See above, p. 80.

111 *Treatise*, p. 52.

112 Modern readers may well be familiar with Ludwig Wittgenstein's discussion of rule-governed behavior and the application of rules in his *Philosophical Investigations*. It is possible that resemblances between the arguments are due to the friendship between the two, but it is hard to say who might have influenced whom, since Wittgenstein came to Cambridge after the first draft but before the publication of the revised version of *Treatise*.

113 *Treatise*, p. 16.

114 Ibid., pp. 15–16.

115 Ibid., pp. 16–17.

116 Ibid., p. 17.

117 A. W. Burks, 'On the significance of Carnap's system of inductive logic for the philosophy of induction,' in Schilpp, op. cit., pp. 739–59.

118 *Treatise*, p. 3.

119 In 'The metaphysical status of P,' above.

120 *LFP*, p. 178.

121 Ibid., p. 181.

122 Idem, emphasis added.

123 I cannot cite the precise location of this quote in Joseph Heller's *Catch-22*, because all three of my copies of that excellent book have been ripped off. If any among you don't grasp the allusion, you have a deprived

cultural background and should take immediate steps to get the book yourself and find out about it.

124 Carnap agrees that this is a result of the definition of c^*, but thinks it acceptable, since instance-confirmation can be used for general laws.

125 R. Carnap, 'Replies and expositions', in Schilpp, op. cit., p. 982.

126 In Section VI of A. Burks, op. cit. Carnap's agreement is at pp. 982–3 of Schilpp, op. cit.

127 In Schilpp, op. cit., p. 981.

128 Ibid., p. 983.

129 *LFP*, p. 250.

130 Ibid., p. 269.

131 Ibid., p. 278.

132 Ibid., pp. 528–30.

133 Ibid., p. 518. See also p. 343.

134 R. Carnap, 'Replies and expositions,' in Schilpp, op. cit., p. 994.

135 As opposed to a pure or garden variety psychologism, which occurs 'where the problems themselves are of an objective nature but the descriptions by which the author intends to give a general characterization of the problems are framed in subjectivist, psychological terms (like "thinking")....,' *LFP*, p. 39.

136 This point is discussed in some detail in the next chapter, 'Relative frequency theories of probability.' Briefly: if 40 per cent of heavy smokers but only 30 per cent of churchgoers die before 60, what is the probability that Smith will die before 60 if he is a churchgoing heavy smoker? No general solution to this problem exists.

137 Since in most cases the series has not yet exhausted itself, we cannot know in advance what the relative frequency in the reference class will be in the end.

138 In the section 'Chief criticisms of relative frequency theories.'

139 J. Kemeny, op. cit., secs V & VI *passim*.

140 See below, pp. 139ff.

141 H. Putnam, 'A definition of degree of confirmation for very rich languages,' pp. 58–62.

142 *LFP*, p. 74.

143 E. Nagel, 'Carnap's theory of induction,' in Schilpp, op. cit., p. 792.

144 I do not wish to claim that 'sentences resemble facts' any more or any less than I wish to claim that vector diagrams resemble natural forces. It is useful representation that is at issue – not structural similarity or aesthetic resemblance.

145 Putnam, radio lecture titled 'Probability and confirmation.'

146 Carnap's discussion of this objection, which I have substantially reproduced here, is in *LFP*, pp. 30–1.

147 A. J. Ayer in Foster and Martin, op. cit., p. 73.

148 It might be best to start with Henry A. Kyburg, 'Recent work in inductive logic,' pp. 249–87, for an overview of the situation and a more extensive bibliography.

149 H. Putnam '"Degree of confirmation" and inductive logic,' in Schilpp, op. cit., and in the radio lecture 'Probability and confirmation.'

150 R. Carnap, 'Replies and expositions,' in Schilpp. op. cit., pp. 983–9
151 *LFP*, p. x.
152 Ibid.
153 R. Weatherford, 'Probability and certainty in C. I. Lewis's epistemology.'
154 Burks, op. cit., p. 746.
155 *LFP*, pp. 182–3.
156 Ibid., p. 183.
157 Hans Reichenbach tries, in *The Theory of Probability*, to develop a system of probability inference based on the notion of truth frequencies, but I find it unsuccessful for reasons to be given in the next chapter.
158 J. Kemeny, op. cit., p. 737.

IV RELATIVE FREQUENCY THEORIES OF PROBABILITY

1 E. Nagel, *Principles of the Theory of Probability*, p. 19. Hereafter cited as *Principles*.
2 This view is shared by R. Carnap, *Logical Foundations of Probability*, 1950 edn, p. 28 (hereafter cited as *LFP*), and H. E. Kyburg, *Probability and Inductive Logic*, p. 52, but H. Reichenbach mentions Poisson and Boole in lieu of Venn in his *The Theory of Probability*, p. 68 (hereafter cited as *TOP*).
3 R. von Mises, *Probability, Statistics, and Truth*, p. 3. Hereafter cited as *PST*.
4 *PST*, p. 4
5 Ibid.
6 Ibid. No italicization or single quotes in original.
7 Ibid., p. ix.
8 Ibid.
9 Ibid., p. 10.
10 Ibid., p. 11.
11 Ibid., p. 12.
12 Ibid.
13 Ibid., p. 15.
14 Ibid., pp. 24–5.
15 Ibid., pp. 39–57, summarized. Von Mises would probably prefer to use probability values other than 1/6, to show that the rules are not based on the Principle of Indifference but apply also to biased dice. I chose to use the more familiar value to facilitate understanding of the operations themselves.
16 *TOP*, pp. 68–9.
17 Ibid., p. 132.
18 Ibid., p. 70.
19 Ibid., p. 52.
20 These are explained on pp. 53–67 of *TOP*.
21 Ibid., pp. 45–6.

22 Ibid., p. 48.
23 See especially *PST*, pp. 31–3.
24 Ibid., p. 32.
25 H. Reichenbach, *Axiomatik der relativistischen Raum-Zeit-Lehre*, Braunschweig, 1924, and *Philosophie der Raum-Zeit-Lehre*, Berlin and Leipzig, 1928.
26 *PST*, pp. 31–2.
27 Ibid., pp. vii, 31.
28 Ibid., p. 30.
29 Ibid., pp. 75, 77, 80.
30 *TOP*, p. 444.
31 Ibid., p. 359, emphasis in original.
32 Ibid.
33 Ibid., p. 364.
34 Ibid., p. 446. The parallelism between Reichenbach's system and the system of Carnap and Lewis is so marked that Carnap has said: 'It seems to me that it would be more in accord with Reichenbach's own analysis if his concept of weight were identified [not with relative frequency but] instead with the estimate of relative frequency. If Reichenbach's theory is modified in this one respect, our conceptions would agree in all fundamental points.' (*LFP*, p. 176.)
35 'A posit is a statement with which we deal as true, although the truth value is unknown.' (*TOP*, p. 373.)
36 *PST*, p. 11.
37 *TOP*, p. 377.
38 G. Boole, *The Laws of Thought*, London, 1854, p. 247.
39 *TOP*, p. 378
40 K. Popper, *Logik der Forschung*.
41 *TOP*, p. 395.
42 Ibid., p. 378.
43 Ibid., pp. 378–9.
44 Ibid., p. 409.
45 *LFP*, p. 1.
46 *TOP*, p. 71.
47 On pp. 17–18 of *PST* he calls such an application 'utter nonsense.'
48 See above, pp. 52–73.
49 *PST*, p. 199.
50 Von Mises sometimes gives the impression that he thinks his condition of randomness prevents the occurrence of very unusual runs or distributions (*PST*, pp. 111–15, for example). It does indeed exclude certain sequences which would otherwise violate the Law of Large Numbers (such as the sequences of pp. 111–12 based on a square root table). But the randomness condition only forbids certain types of *infinite* sub-sequences. If the first million runs of the scintillation experiment gave the frequency 0.999, and the remaining infinite repetitions gave 0.211, the series would approach 0.211 as a limit and would remain random according to the definition, because any infinite ('extended indefinitely') sub-sequence which includes all of the first million experi-

ments will find those 0.999 values swamped by another million, then a quadrillion, then a decillion, then an infinity of repetitions giving 0.211 as the frequency. It follows, then, that no extreme of 'strangeness' in any *finite* sub-sequence is forbidden by the condition of randomness.

51 *TOP*, p. 106.
52 The argument appears on pp. 17–18 of *PST*.
53 *PST*, p. 19.
54 J. M. Keynes, *A Treatise on Probability*, pp. 130–1. Hereafter cited as *Treatise*.
55 On p. 71 of *TOP*, for example, he says that traditional mathematical talk about 'the relative probability of *C* with respect to *B*... does not seem advisable because all probabilities are relative....'
56 *TOP*, p. 375.
57 *PST*, p. 17
58 Ibid., pp. 177–83.
59 Ibid., pp. 202–20.
60 I have argued elsewhere (in my Bechtel Prize Essay, 'The Heisenberg uncertainty relations'), that quantum mechanics *does not* imply this.
61 Although many feel that Abraham Wald's 'Die Widerspruchsfreiheit des Kollektivbegriffs der Wahrscheinlichkeitsrechnung' [*Ergebnisse math. Kolloquium*, 8(1937), pp. 38–72] obviates the worst difficulties with the concept, it is not generally considered essential.
62 *PST*, pp. vii, 31.
63 This is what it means if one accepts the correspondence theory of truth, or Tarski's semantic concept. For a pragmatist, the truth of *P* would mean that we could count on future experience being such that 'acting on' *P* would lead to 'success in practice.' I would like to say what it means for someone who accepts the coherence theory of truth, but I have never understood how such a theory can escape solipsism. Such considerations as these properly arise when one uses our simplistic method (What are the metaphysical implications of *P*? Well, what must reality be like if *P* is *true*?). It could well be argued that passing from problems of meaning to problems of truth need not constitute an advance. Our interest, however, is not so much in arriving at metaphysical knowledge as it is in understanding probability concepts and their metaphysical implications. For this I think a rough and ready correspondence theory of truth should suffice.
64 *PST*, p. 84.
65 Ibid.
66 Ibid., p. 85.
67 Ibid., p. 14. Strictly speaking this statement is false even in von Mises' system, for a probability is not a property of a die or any other material object. It is a property of a collective of (in this case) repetitive simple events, the castings of this die. Nevertheless, it is clear that the probability is an empirical quality which is based on physical facts (in particular, it is based upon the physical constitution of this die, and, to a lesser extent, the environment in which it is rolled) and is manifested in the physical universe.

68 This discussion is primarily based on sections 66 and 67 of *TOP*.

69 Actually 8/47, since I have already seen 5 cards which are neither Aces nor Tens, but we will ignore this refinement.

70 *PST*, p. 31.

71 Ibid., p. 33.

72 Ibid., p. 32.

73 Ibid., p. 65.

74 *LFP*, p. 34. The Waismann paper is 'Logische Analyse des Wahrscheinlichkeitsbegriffs,' *Erkenntnis*, I, 1930–1, pp. 228–48.

75 *PST*, p. 16.

76 Ibid., p. 142.

77 This is explicitly recognized on p. 103 of *PST*. Nagel notes that *no special axioms* are required; an RF theory can be deduced from any axioms adequate for real number theory (including limits). See *Principles*, p. 38.

78 *TOP*, p. 350.

79 Ibid., p. 3.

80 Ibid., p. 12.

81 Ibid., p. 52.

82 Ibid., pp. 343–4.

83 Ibid., p. 375.

84 Ibid., p. 70.

85 Ibid., pp. 429–33.

86 Ibid., p. 446.

87 Ibid., p. 460.

88 Much time and effort has been wasted by contemporary scholars in an unfortunate debate about whether or not this is true. Critics of the RF theory have pressed the point that, for any N, however large, it is always possible that the series might begin to diverge after N, and, since there remains an infinite portion of the series, this divergence can always outweigh the finite segment up to N, so that the relative frequency will diverge from P as much as you like. Supporters of the RF position respond that they do not mean you can *pick* an N which marks a convergence point, only that if the series converges to a limit it is logically necessary that there *be* a convergence point.

Here I think the supporters are clearly right in their mathematical logic, but the attackers succeed in making the pragmatic point that we can't be sure of attaining convergence in any finite stretch of experience, however long it might be, even if we know a limit exists.

89 *TOP*, p. 479.

90 H. Reichenbach, 'On the justification of induction,' p. 98.

91 This justification of RF statements as scientific is particularly clear in Carnap, *LFP*, p. 501, but it is also present throughout Reichenbach's *TOP* and Nagel's *Principles*.

92 Students of Hilary Putnam will realize that I am ignoring the difficulties that he (and others, such as Carl Hempel) raised about Auxiliary Statements and their role in science. In this case, however, I think we can safely conflate AS with theoretical statements, without any significant logical error.

93 Here again I treat disruptive forces, over-simplifications, and other of Putnam's auxiliary statements as examples of theoretical error, though I quite agree they are not in general the same.

94 Von Mises has an interesting but unpersuasive argument to the effect that *all* physical properties are therefore equivalent to the limit of a relative frequency in an infinite series. See the last chapter of *PST*.

95 It is explicitly mentioned by T. L. Fine, *Theories of Probability*, p. 103; Kyburg, op. cit., p. 45; A. J. Ayer, *Probability and Evidence*, p. 48; and even Reichenbach, *TOP*, p. 352.

96 For the difference between intensionally and extensionally given sequences, see *TOP*, pp. 339–40.

97 C. I. Lewis, *An Analysis of Knowledge and Valuation*, 1962 edn, p. 283. Hereafter *AKV*.

98 Discussions of this or parallel points occur in W. Kneale, *Probability and Induction*, p. 42; Ayer, op. cit., pp. 47–50; Nagel, *Principles*, p. 52; and Lewis, *AKV*, p. 288.

99 *TOP*, p. 343. A similar statement appears on p. 344.

100 In the section entitled 'The epistemological status of P.'

101 *AKV*, p. 289.

102 Reichenbach of course holds that this is not necessary since *P* is not asserted as probable nor even as true but as a posit justified by the Rule of Induction.

103 This is commonly attributed to Sir Karl Popper, but he himself insisted that he understood falsifiability as a criterion of demarcation between science and non-science, rather than a standard of meaningfulness. The misconstrual has none the less become a standard alternative to the Logical Positivists' verification criterion of meaning.

104 Modus tollens is a valid form of deductive inference:

$$P \text{ implies } Q$$

$$Q \text{ is false}$$

$$\text{Therefore } P \text{ is also false.}$$

105 *Treatise*, p. 95.

106 See above, pp. 165–6.

107 For details, see Fine, op. cit., pp. 93–102; Keynes, *Treatise*, pp. 102–3, 167; Ayer, op. cit., pp. 44–7, 51, etc.

108 Arguments of this general type, or related criticisms, can be found in Fine, op. cit., pp. 103 and 239; Kneale, op. cit., p. 165; and Kyburg, op. cit., p. 50.

109 Kneale, op. cit., p. 194.

110 *LFP*, p. 176. Other criticisms of RF theory are also raised in this section.

111 See L. Wittgenstein, *Philosophical Investigations*, for this distinction and the complex relationships between criteria, concepts, and meanings.

112 See above, pp. 149ff., and *LFP*, p. 34.

113 Fine, op. cit., p. 103.

114 Reichenbach describes and praises this movement in *The Rise of Scientific Philosophy*.

V   THE SUBJECTIVISTIC THEORY OF PROBABILITY

1 See, for example, E. Nagel, *Principles of the Theory of Probability*, *passim*.
2 In chapter II, 'The classical theory of probability.'
3 F. Ramsey, 'Truth and probability,' in *The Foundations of Mathematics and Other Logical Essays*.
4 Ibid, pp. 256–7.
5 D. Davidson, P. Suppes, and S. Siegel, *Decision Making: An Experimental Approach*.
6 B. de Finetti, *Theory of Probability: A Critical Introductory Treatment*, p. x. Emphasis in original. Hereafter cited as *Theory*.
7 Ramsey, op. cit., p. 1.
8 B. de Finetti, *Probability, Induction, and Statistics: The Art of Guessing*, p. xiv.
9 Ramsey, op. cit., p. 167.
10 B. de Finetti, 'Foresight: Its Logical Laws, Its Subjective Sources,' p. 104. Hereafter cited as *Foresight*.
11 Ibid., p. 103.
12 Even our own death (life-insurance), but not, perhaps, the immediate and absolute end of the universe.
13 Davidson, Suppes, and Siegel, op. cit., p. 75. The authors also cite similar findings in another team's experiments, and contrary evidence in a third investigation. (A four-sided die has two ends rounded off, so that only four sides can 'come up.')
14 Ibid., p. 54.
15 *Foresight*, p. 101.
16 L. Savage, 'The foundations of statistics reconsidered,' 1964 rep., p. 180. That this is true can be seen by reflecting that 'laws' of statistics are deductive consequences of the axioms of the probability calculus, which we have already said are equivalent to the coherency condition.
17 *Foresight*, p. 116.
18 H. Jeffreys, *Scientific Inference*, p. 34.
19 *Foresight*, p. 115.
20 De Finetti actually uses the outcome of an election as one of his examples, *Theory*, p. 59.
21 Charles L. Stevenson, *Ethics and Language*.
22 *Theory*, p. 218.
23 De Finetti, *Probability, Induction, and Statistics*, p. 154.
24 Approximate rendition from memory of Brother Dave Gardner.
25 *Foresight*, p. 142.
26 De Finetti, *Probability, Induction, and Statistics*, p. 229.
27 Ibid., p. 212.
28 E. Borel, 'Apropos of a Treatise on Probability,' in Kyburg and Smokler, op. cit., p. 50.
29 Idem.
30 De Finetti, *Probability, Induction and Statistics*, p. vi.
31 *Theory*, p. 6, paraphrased.

32 Ibid., p. 218.
33 *Foresight*, pp. 149, 154.
34 I. J. Good, 'Subjective probability as the measure of a non-measurable set,' p. 319.
35 Ramsey, op. cit., p. 157.
36 *Theory*, ch. 5.
37 Ramsey, op. cit., p. 196.
38 *Theory*, p. 72.
39 The mistake of psychological hedonism lies not in this principle, which is tautological, but in the principle that only pleasure is desirable, which is false.
40 Nor did they intend to do so. They clearly and explicitly noted that their investigations were independent of and had no bearing on questions of objective probability (op. cit., p. 11).
41 L. J. Savage, *The Foundations of Statistics*, pp. 61–2.
42 This criticism appears also in the chapter on RF theories, above.
43 R. Carnap, *Logical Foundations of Probability*, 1950 edn, p. 51.

VI   CONCLUSION

1 W. V. O. Quine, 'Foundations of mathematics,' p. 24.
2 N. Goodman, *Fact, Fiction, and Forecast*, p. 66.
3 R. Carnap, *Logical Foundations of Probability*, 1950 edn, p. 1.
4 H. Putnam, 'What is mathematical truth?,' pp. 69–70.
5 Ibid., pp. 73–4.
6 '[There is an] essential duality of probability, which is both epistemic and aleatory. Aleatory probabilities have to do with the physical state of coins or mortal humans. Epistemic probabilities concern our knowledge.' I. Hacking, *The Emergence of Probability*, pp. 122–3.
7 H. Putnam, 'The analytic and the synthetic,' pp. 378–9.
8 For my views on this, see my Bechtel Prize Essay, 'The Heisenberg uncertainty relations.'
9 For the distinction between a general concept and more specific conceptions, see J. Rawls, *A Theory of Justice*, pp. 5–6.
10 I think the 'likelihood' approach to probability theory now being developed has the virtue of emphasizing the long-neglected problem of physical chance by its attempts to tie probability to actual experimental chance set-ups. I have not covered that view because it seems to me to be quite unsuccessful in what it is attempting to do and it is in any case too new and undeveloped to count as a major theory – perhaps in a later edition it will warrant inclusion. In any case I'm glad the problem is being addressed.
11 H. Putnam, 'What is mathematical truth?,' p. 66.

BIBLIOGRAPHY

Adler, H. L., and Roessler, E. B., *Introduction to Probability and Statistics*, 3rd edn, San Francisco, W. H. Freeman, 1964.

Arley, N., and Buch, K. R., *Introduction to the Theory of Probability and Statistics*, New York, John Wiley, 1950.

Ayer, A. J., 'The conception of probability as a logical relation,' in S. Korner (ed.), *Observation and Interpretation*, 1957. (Reprinted in M. H. Foster and M. L. Martin (eds), *Probability, Confirmation, and Simplicity*, 1966.)

Ayer, A. J., *Probability and Evidence*, New York, Columbia University Press, 1972.

Bayes, T., 'An essay toward solving a problem in the doctrine of chances,' with R. Price's foreword and discussion, *Philosophical Transaction of the Royal Society*, pp. 370–418, 1763, reprinted in W. E. Deming (ed.), *Facsimiles of Two Papers by Bayes*, New York, Hafner, 1963.

Bernoulli, D., 'The most probable choice between several discrepant observations and the formation therefrom of the most likely induction,' *Acta Acad. Petrop.*, 1777, trans. C. G. Allen, in E. S. Pearson and M. G. Kendall (eds), *Studies in the History of Statistics and Probability*, 1970.

Bernoulli, J., *Ars conjectandi*, excerpted in J. R. Newman (ed.), *The World of Mathematics*, New York, Simon & Schuster, 1956, and in F. Maseres, *The Doctrine of Chances*, 1795.

Black, M., 'Induction and probability,' in R. Klibansky (ed.), *Philosophy in the Mid-century*, Florence, 1958.

Black, M., 'Probability,' *Encyclopedia of Philosophy*, New York, Macmillan and Free Press, 1967.

Borel, E., 'Apropos of a Treatise on Probability,' *Revue Philosophique*, 1924, trans. H. E. Smokler in H. E. Kyburg Jr, and H. E. Smokler (eds), *Studies in Subjective Probability*, 1964.

Borel, E., *Probability and Certainty*, New York, Walker, 1963.

Borel, E., *Probabilities and Life*, trans. Maurice Baudin, New York, Dover, 1962 (published in 1953 as *Les Probabilitís et la Vie*).

Born, M., 'Einstein's statistical theories,' in P. A. Schilpp (ed.), *Albert Einstein: Philosopher-Scientist*, New York, Harper Torchbooks, 1959.

273

Bibliography

Braithwaite, R. B., 'On unknown probabilities,' in S. Korner (ed.), *Observation and Interpretation*, 1957.

Broad, C. D., 'Critical notice: *Wahrscheinlichkeit, Statistik und Wahrheit*,' *Mind*, 46, 1937.

Burks, A. W., 'On the significance of Carnap's system of inductive logic for the philosophy of induction,' in P. A. Schilpp (ed.), *The Philosophy of Rudolf Carnap*, 1963.

Carnap, R., 'The aim of inductive logic,' in E. Nagel, P. Suppes, and A. Tarski (eds), *Logic, Methodology and Philosophy of Science*, 1962.

Carnap, R., *The Continuum of Inductive Methods*, University of Chicago Press, 1952.

Carnap, R., *Logical Foundations of Probability*, University of Chicago Press, 1950.

Carnap, R., 'My basic conceptions of probability and induction,' in P. A. Schilpp (ed.), *The Philosophy of Rudolf Carnap*, 1963.

Carnap, R., 'On inductive logic,' *Philosophy of Science*, vol. 12, no. 2, pp. 72–97, April 1945. (Reprinted in M. H. Foster and M. L. Martin (eds), *Probability, Confirmation, and Simplicity*, 1966.)

Carnap, R., 'Replies and expositions,' in P. A. Schilpp (ed.), *The Philosophy of Rudolf Carnap*, 1963.

Carnap, R., 'The two concepts of probability,' *Philosophy and Phenomenological Research*, 5 (1945), pp. 513–32. (Reprinted in H. Feigl and W. Sellars (eds), *Readings in Philosophical Analysis*, New York, Appleton-Century-Crofts, 1949.)

Carnap, R., and Jeffrey, R. C., *Studies in Inductive Logic and Probability*, Berkeley, University of California Press, vol. I, 1971 (posthumous for Carnap, who died in September 1970).

Cheng, Chung-ying, 'On Peirce's and Lewis' theories of induction,' Doctoral Thesis, Harvard University Archives, 1963.

David, F. N., *Games, Gods and Gambling*, New York, Hafner, 1962.

Davidson, D., Suppes, P., and Siegel, S., *Decision Making: An Experimental Approach*, Stanford University Press, 1957.

Feigl, H., 'De principiis non disputandum...?,' in M. Black (ed.), *Philosophical Analysis*, Englewood Cliffs, Prentice-Hall, 1963.

Feigl, H., and Maxwell, G. (eds), *Current Issues in the Philosophy of Science*, New York, Holt, Rinehart & Winston, 1961.

Feller, W., *An Introduction to Probability Theory and Its Applications*, New York, John Wiley, 1970.

Fine, T. L., *Theories of Probability*, New York, Academic Press, 1973.

Finetti, B. de. 'Foresight: Its Logical Laws, Its Subjective Sources,' *Annales de l'Institut Henri Poincaré*, vol. 7, 1937, translated by Henry Kyburg in H. E. Kyburg Jr and H. E. Smokler (eds), *Studies in Subjective Probability*, 1964.

Finetti, B. de, *Probability, Induction, and Statistics: The Art of Guessing*, New York, John Wiley, 1972.

Finetti, B. de. *Theory of Probability: A Critical Introductory Treatment*, vol. I, trans. Antonio Machi and Adrian Smith, New York, John Wiley, 1974. Originally *Teoria della Probabilitá*, 1970.

Foster, M. H., and Martin, M. L. (eds), *Probability, Confirmation, and Simplicity*, New York, Odyssey Press, 1966.

Frechet, M., 'The diverse definitions of probability,' *Journal of Unified Science*, 8, 1939–40.

Friedman, M., 'Truth and confirmation,' *Journal of Philosophy*, July 1979.

Good, I. J., *The Estimation of Probabilities: An Essay on Modern Bayesian Methods*, Cambridge, Mass., Research Monograph no. 30 of the MIT Press, 1965.

Good, I. J., 'Subjective probability as the measure of a non-measurable set,' in E. Nagel, P. Suppes, and A. Tarski (eds), *Logic, Methodology, and Philosophy of Science*, 1962.

Goodman, N., *Fact, Fiction, and Forecast*, New York, Bobbs–Merrill, 1955.

Goodstein, R. L., 'On von Mises' theory of probability,' *Mind*, 49, 1940.

Hacking, I., *The Emergence of Probability*, Cambridge University Press, 1975.

Hacking, I. *The Logic of Statistical Inference*, Cambridge University Press, 1965.

Hacking, I. 'One problem about induction,' in Imre Lakatos (ed.), *The Problem of Inductive Logic*, 1968.

Hempel, C. G., 'Rudolf Carnap,' *Collier's Encyclopedia*, vol. 5, New York, Crowell–Collier & Macmillan, 1956.

Hempel, C. G., 'Studies in the logic of confirmation,' *Mind*, 54, 1945.

Hodges, J. and Lehmann, E., *Basic Concepts of Probability and Statistics*, San Francisco, Holden-Day, 1964.

Hogben, L., *Mathematics in the Making*, London, Rathbone Books, 1960.

Jeffrey, R., *The Logic of Decision*, New York, McGraw-Hill, 1965.

Jeffreys, H., *Scientific Inference*, Cambridge University Press, 1937, p. 34.

Kemeny, J., 'Carnap on probability and induction,' in P. A. Schilpp (ed.), *The Philosophy of Rudolf Carnap*, 1963.

Kemeny, J., 'Fair bets and inductive probabilities,' *Journal of Symbolic Logic*, 20, 1955.

Kendall, M., 'The beginnings of a probability calculus,' in E. S. Pearson and M. G. Kendall (eds), *Studies in the History of Statistics and Probability*, 1970.

Keynes, John Maynard, *Essays in Biography*, New York, W. W. Norton, 1963. Originally published in London, 1933.

Keynes, John Maynard, *A Treatise on Probability*, London, Macmillan, 1921; New York, Harper & Row, 1962.

Keynes, John Maynard, *Two Memoirs*, New York, Augustus M. Kelley, 1949.

Kneale, W., *Probability and Induction*, Oxford University Press, 1949.

Kolmogorov, A., *Foundations of the Theory of Probability*, New York, Chelsea, 1951.

Korner, S. (ed.), *Observation and Interpretation* (commonly known as 'The Colston Papers'), Proceedings of the Ninth Symposium of the Colston Research Society, London, Butterworths Scientific Publications, 1957.

Kyburg Jr, H. E., *Probability and Inductive Logic*, London, Macmillan, 1970.

Kyburg Jr, H. E., *Probability and the Logic of Rational Belief*, Middletown, Conn., Wesleyan University Press, 1961.

Kyburg Jr, H. E., 'Recent work in inductive logic,' *American Philosophical Quarterly*, 1, 1964, pp. 249–87.

Kyburg Jr, H. E., and Nagel, E. (eds), *Induction: Some Current Issues*, Middletown, Conn., Wesleyan University Press, 1963.

Kyburg Jr, H. E., and Smokler, H. E. (eds), *Studies in Subjective Probability*, New York, John Wiley, 1964.

Lakatos, Imre (ed.), *The Problem of Inductive Logic*, Amsterdam, North-Holland, 1968.

Laplace, P. S. de, *A Philosophical Essay on Probability*, New York, Dover, 1951. First published in 1820.

Leblanc, H., 'Statistical and Inductive Probabilities,' in H. E. Kyburg Jr and E. Nagel (eds), *Induction: Some Current Issues*, 1963.

Lenz, J., 'Carnap on defining "degree of confirmation",' *Philosophy of Science*, 23 (2), July 1956, pp. 230–6. (Reprinted in M. H. Foster and M. L. Martin (eds), *Probability, Confirmation, and Simplicity*, pp. 73–9.)

Lenz, J., 'The frequency theory of probability,' in E. Madden (ed.), *The Structure of Scientific Thought*, Boston, Houghton Mifflin, 1960. (Reprinted in M. H. Foster and M. L. Martin (eds), *Probability, Confirmation, and Simplicity*, 1966.)

Levi, I., *Gambling With Truth*, New York, Knopf, 1967.

Lewis, C. I., *An Analysis of Knowledge and Valuation*, La Salle, Ill., Open Court Press, 1962 (originally 1946).

Lewis, C. I., 'Review of John Maynard Keynes's *A Treatise on Probability*,' *The Philosophical Review*, XXXI (2), 1922.

Mackie, J., *Truth, Probability, and Paradox*, Oxford, Clarendon Press, 1973.

Maistrov, L. E., *Probability Theory: A Historical Sketch*, trans. S. Kotz, New York, Academic Press, 1974.

Margenau, H., 'On the frequency theory of probability,' *Philosophy and Phenomenological Research*, 6, 1945–6.

Marks, R., *The New Mathematics Dictionary and Handbook*, New York, Bantam Books, 1964.

Martin, N. M., 'Rudolf Carnap,' *Encyclopedia of Philosophy*, vol. 2, New York, Macmillan and Free Press, 1967.

Maseres, F., Esquire, Cursitor Baron of the Court of Exchequer, *The Doctrine of Permutations and Combinations, Being an Essential and Fundamental Part of the Doctrine of Chances* (short title: *The Doctrine of Chances*), London, B. & J. White, 1795.

Mehlberg, J., 'Is a unitary approach to foundations of probability possible?,' in H. Feigl and G. Maxwell (eds), *Current Issues in the Philosophy of Science*, 1961.

Mises, R. von, *Mathematical Theory of Probability and Statistics*, ed. H. Geiringer, New York and London, Academic Press, 1964.

Mises, R. von, *Probability, Statistics, and Truth*, London, George Allen & Unwin, Macmillan, 1957. Originally published as *Wahrscheinlichkeit, Statistik und Wahrhelt, Einführung in die neue Wahrscheinlichkeitslehre, und ihre Anwendung*, Vienna, 1928.

Morgenbesser, S., Suppes, P., and White, M. (eds), *Philosophy, Science, and*

Method: *Essays in Honor of Ernest Nagel*, New York, St Martin's Press, 1969.

Mosteller, F., Rourke, R., and Thomas, G. (eds), *Probability with Statistical Applications*, Reading, Mass., Addison-Wesley, 1970 (1st edn 1961).

Nagel, E., 'Carnap's theory of induction,' in P. A. Schilpp (ed.), *The Philosophy of Rudolf Carnap*, 1963.

Nagel, E., *Principles of the Theory of Probability*, vol. 1, no. 6 of *International Encyclopedia of Unified Science*, University of Chicago Press, 1939.

Nagel, E., Suppes, P., and Tarski, A. (eds), *Logic, Methodology and Philosophy of Science*, Stanford University Press, 1962.

Newman, J. R., 'Commentary on the law of large numbers,' in J. Newman (ed.), *The World of Mathematics*, 1956.

Newman, J. R. (ed.), *The World of Mathematics*, New York, Simon & Schuster, 1956.

Pearson, E. S., and Kendall, M. G. (eds), *Studies in the History of Statistics and Probability*, London, Charles Griffin, 1970.

Popper, K., *Conjectures and Refutations*, London, Routledge & Kegan Paul, 1963.

Popper, K., *The Logic of Scientific Discovery*, New York, Harper & Row, 2nd edn, 1968, originally published as *Logik der Forschung*, Vienna, 1934 (though imprinted '1935').

Putnam, H., 'The analytic and the synthetic,' *Minnesota Studies in the Philosophy of Science*, vol. III, Minneapolis, University of Minnesota Press, 1962.

Putnam, H., 'A definition of degree of confirmation for very rich languages,' *Philosophy of Science*, XXIII, 1956.

Putnam, H., '"Degree of confirmation" and inductive logic,' in P. A. Schilpp (ed.), *The Philosophy of Rudolf Carnap*, 1963.

Putnam, H., *Mathematics, Matter and Method, Philosophical Papers*, vol. 1, Cambridge University Press, 1975.

Putnam, H., 'Probability and confirmation' (radio lecture), *Voice of America*, Forum Philosophy of Science, 10, US Information Agency, 1963.

Putnam, H., 'What is mathematical truth?,' in H. Putnam, *Mathematics, Matter and Method, Philosophical Papers*, vol. 1, Cambridge University Press, 1975.

Quine, W. V. O., 'Foundations of mathematics,' collected in *The Ways of Paradox*, New York, Random House, 1966.

Ramsey, F., *The Foundations of Mathematics and Other Logical Essays*, London, Kegan Paul, 1931.

Rawls, J., *A Theory of Justice*, Cambridge, Mass., Harvard University Press, 1971.

Reichenbach, H., 'On the justification of induction,' *Journal of Philosophy*, 37, 1940.

Reichenbach, H., *The Rise of Scientific Philosophy*, Berkeley, University of California Press, 1951.

Reichenbach, H., *The Theory of Probability*, Berkeley, University of California Press, 1949, the author's translation (with new additions) of his *Wahrscheinlichkeitslehre*, Leiden, 1935.

Bibliography

Salmon, W., 'The justification of inductive rules of inference,' in I. Lakatos (ed.), *The Problem of Inductive Logic*, 1968.

Savage, L., *The Foundations of Statistics*, New York, Dover, 1972 (and New York, John Wiley, 1954).

Savage, L., 'The foundations of statistics reconsidered,' *Proceedings of the Fourth Berkeley Symposium on Mathematics and Probability*, Berkeley, University of California Press, 1961. (Reprinted in H. E. Kyburg Jr and H. E. Smokler (eds), *Studies in Subjective Probability*, 1964.)

Schilpp, P. A. (ed.), *The Philosophy of C. I. Lewis*, La Salle, Ill., Open Court Press, 1968.

Schilpp, P. A. (ed.), *The Philosophy of Rudolf Carnap*, La Salle, Ill., Open Court Press, 1963.

Skyrms, B., *Choice and Chance*, Belmont, Cal., Dickenson, 1966.

Stevenson, C. L., *Ethics and Language*, London, Oxford University Press, 1944.

Todhunter, I., *A History of the Mathematical Theory of Probability From the Time of Pascal to That of Laplace*, New York, Chelsea, 1949 (first published in London, 1865).

Uvarov, E. B., Chapman, D. R., and Isaacs, A., *A Dictionary of Science*, Baltimore, Penguin Books, 1943, revised 1964.

Venn, J., *The Logic of Chance*, Macmillan, London, 1866 (reprinted New York, Chelsea, 1963).

Wald, A., *Statistical Decisions Functions*, New York, John Wiley, 1950; 2nd edn, New York, Chelsea, 1971.

Weatherford, R., 'The Heisenberg uncertainty relations,' Bechtel Prize Essay, Harvard University Archives, 1970.

Weatherford, R., 'Probability and certainty in C. I. Lewis's epistemology,' doctoral thesis, Harvard University Archives, 1972.

Whittle, P., *Probability*, London, John Wiley, 1976 (first published by Penguin, 1970).

Wisdom, J., 'A note on probability,' in M. Black (ed.), *Philosophical Analysis*, Englewood Cliffs, N. J., Prentice-Hall, 1963.

Wittgenstein, L., *Philosophical Investigations*, trans. G. E. M. Anscombe, New York, Macmillan, 1953.

Wittgenstein, L., *Tractatus Logico-Philosophicus*, trans. D. F. Pears and B. F. McGuinness, London, Routledge & Kegan Paul, 1961 (written in 1921).

Wright, G. H. von, *The Logical Problem of Induction*, Helsinki, *Acta Philosophica Fennica*, fasc. 3, 1941; 2nd rev. edn, Oxford, Blackwell, 1957.

Wright, G. H. von, 'Remarks on the epistemology of subjective probability,' in E. Nagel, P. Suppes, and A. Tarski (eds), *Logic, Methodology and Philosophy of Science*, 1962.

Wright, G. H. von, *A Treatise on Induction and Probability*, New York, Harcourt, Brace & World, 1951.

INDEX

absolute probability: on a priori theory, 108–12; on Classical theory, 41; on Relative Frequency theory, 173–8; on Subjectivistic theory, 233–4

Alembert, Jean d', 25–6

analytic statements, 8, 9, 12, 60, 84, 112, 122ff., 184–7, 191–3

A priori theory of probability, *see* Probability, a priori theory of

Aristotle, 13, 144

Ars Conjectandi, 22–3

Astragalus, 19

Augustus, Caesar, 19

Ayer, A. J., 69, 105, 111–12, 138

Bar-Hillel, 139

Bayes's Postulate, 24, 34

Bayes's Theorem, 7, 24–5, 32–4, 36, 38–40, 68–71, 109–12, 117, 214, 220 225–6, 230, 241

Bayes, Thomas, 24–5, 32ff., 60, 71, 74, 78

Bernoulli, Daniel, 22–3, 25, 31–2, 50–1

Bernoulli, James (Jakob, Jacques), 21, 22–3, 27ff., 78, 86, 111, 219

Bernoulli, John (Johann, Jean), 22–3

Bernoulli, Nicholas (Nikolaus, Nicolas), 22–3

Bernoulli's Limit Theorem, 38–9, 61

Bernoulli's Principle of Fluid Dynamics, 22

Bernoulli's Theorem, 7, 22, 37–8, 61–6, 109, 170, 211, 214, 230, 232; inverse of, 7, 61–6, 68–71

Bertrand's Box, 55–6

Bertrand's Paradox, 56

Black, Max, 15–16, 17

Boltzmann, Ludwig, 71

Bolzano, 13, 144

Boole, George, 162

Borel, Emile, 26, 233

Born, Max, 72–3, 252

Bose–Einstein statistics, 71–3, 74, 251–2

Braithwaite, R. B., 6

Buffon, G. L., 26

Burks, A. W., 120–1, 125, 141

Calcagnini, Celio, 20

Calculus of probability, *see* Probability, Calculus of

Cardano, Girolamo, 20, 21, 27, 29, 51

Carnap, Rudolf, 6, 8, 12, 13, 14, 16, 17, 27, 28, 49, 67–8, *75–143*, 147, 153, 159, 164–5, 166, 189, 205, 206, 209, 211, 213–17, 242, 245–6, 248, 250, 251

Castor, Betty, 240

Catch-22, 123–4

Classical theory of probability, *see* Probability, classical theory of

Claudius, Emperor of Rome, 20

Collectives, 148–50, 160, 174, 188

Completeness, requirement of, 134–5

Confirmation-functions (c-functions), 86ff., 198; 'best possible' c-functions, 139ff.

Cournot, Antoine, 13, 43, 106, 144

International Library of Philosophy

Editor: Ted Honderich

(Demy 8vo)